TJ 809.95 .S68 1988 Vol. 1

NEW ENGLAND INSTITUTE
OF TECHNOLOGY
LEARNING RESOURCES CENTER

History and Overview of Solar Heat Technologies

Solar Heat Technologies: Fundamentals and Applications
Charles A. Bankston, editor-in-chief

1. *History and Overview of Solar Heat Technologies*
Donald A. Beattie, editor

2. *Solar Resources*
Roland L. Hulstrom, editor

3. *Economic Analysis of Solar Thermal Energy Systems*
Ronald E. West and Frank Kreith, editors

4. *Fundamentals of Building Energy Dynamics*
Bruce D. Hunn, editor

5. *Solar Collectors, Energy Storage, and Materials*
Francis de Winter, editor

6. *Active Solar Systems*
George Löf, editor

7. *Passive Solar Buildings*
J. Douglas Balcomb, editor

8. *Passive Cooling*
Jeffrey Cook, editor

9. *Solar Building Architecture*
Bruce Anderson, editor

10. *Implementation of Solar Thermal Technology*
Ronal Larson and Ronald E. West, editors

History and Overview of Solar Heat Technologies

edited by Donald A. Beattie

The MIT Press
Cambridge, Massachusetts
London, England

©1997 Massachusetts Institute of Technology

All rights reserved. No part of this book may be reproduced in any form or by any electronic or mechanical means (including photocopying, recording, or information storage and retrieval) without permission in writing from the publisher.

This book was set in Times Roman by Asco Trade Typesetting Ltd., Hong Kong and was printed and bound in the United States of America.

Library of Congress Cataloging-in-Publication Data

History and overview of solar heat technologies / edited by Donald A. Beattie.
   p. cm. — (Solar heat technologies; 1)
 Includes bibliographical references and index.
 ISBN 0-262-02415-2 (alk. paper)
 1. Solar energy. 2. Solar energy—History. 3. Solar heating.
4. Solar heating—History. I. Beattie, Donald A. II. Series.
TJ809.95.S68 1988 vol. 1
[TJ810]
697′.78s—dc20
[621.47]   96-34297
                                                                 CIP

Contents

	Preface and Acknowledgments Donald A. Beattie, Frederick H. Morse	vii
	Series Acknowledgments Charles A. Bankston	xi
	List of Acronyms	xiii
1	**About this Volume** Charles A. Bankston and Donald A. Beattie	1
2	**About this Series** Charles A. Bankston	15
3	**The Early Years of Federal Solar Energy Programs** Donald A. Beattie	23
4	**The Buildup Years: 1974–1977** Donald A. Beattie	45
5	**The Growth Years: 1977–1980** Frederick H. Morse	109
6	**The Contraction and Redirection Years: 1981–1988 (and Beyond)** Frederick H. Morse	139
7	**Significant Results of the Federal Solar R&D Programs** Charles A. Bankston	181
8	**Observations and Lessons Learned** Charles A. Bankston, Donald A. Beattie, and Frederick H. Morse	237
	Appendix: Outline of the Solar Heat Technologies Series	257
	Contributors	265
	Index	269

Preface and Acknowledgments

Although the Solar Heat Technologies documentation project was initiated in 1982, the current volume was not begun until 1994, when the other nine volumes in the series had either been published or were in the process of final editing. This unforeseen late start gave my two coauthors and me the benefit of drawing on the experience and insights of the many contributors to the other volumes.

Focusing on the political, institutional, and programmatic background of this major government research and development effort, we have attempted to provide a broad overview of the solar program, with emphasis on solar heat technologies, to encapsulate its results, and to draw some lessons for similar future programs. Launched as a more or less classical government research and development program, it eventually included two new elements—demonstration and commercialization—that would become highly controversial. Frequent administration and leadership changes combined with adversarial congressional oversight to keep the program in a state of constant turmoil. Nevertheless, an ambitious program evolved, marked by significant advances in solar technology. It is this story, and more importantly the documentation of the knowledge gained, that we believe will be an important and lasting legacy upon which future researchers and managers can build.

This volume would not exist were it not for the dedication of my two collaborators. Fred Morse, who had the vision to start this series, and Chuck Bankston, our patient editor-in-chief, bring to the volume their unique vantage points developed over years of guiding others. I only wish we could have included transcripts of our innumerable lively debates on the meaning of what we were writing. Despite many differences of opinion, we remain colleagues and friends.

Larry Cohen, now editor-in-chief at MIT Press, may be the longest-suffering individual in the publishing business. Ten years and counting, to the benefit of all, he has persevered, and without his steadfastness, the project would have foundered. Larry, we thank you.

I also thank Annie Harris, whose word processing skills and ability to decipher the undecipherable, while at the same time keeping track of innumerable drafts, made our task so much easier.

Several former colleagues reviewed the manuscript and provided valuable comments. Lloyd Herwig, Harold Horowitz, and Harry Johnson not only read and commented on the text but made available personal files containing background material that was invaluable in reconstructing events going back twenty years. John Teem, my former boss and a key player in the early evolution of the solar energy program, also reviewed parts of the manuscript. To all of them, my heartfelt thanks.

Don Kornreich, another former colleague, now at the Midwest Research Institute, searched his files to provide voluminous original source material pertaining to the Solar Energy Research Institute and the regional solar energy centers. I owe my thanks to his efforts because they assured that these important institutes were accurately portrayed.

And to my many, unnamed, former colleagues and co-workers at the National Science Foundation, Energy Research and Development Administration, Department of Energy, and National Aeronautics and Space Administration who labored long and hard to advance solar heat technologies, my deep gratitude for their support through many tumultuous years. I hope that in these pages and in the other volumes of this series, they will recognize some of their own achievements.

—Donald A. Beattie

Writing chapters 5 and 6 was a challenging undertaking that began with lengthy interviews with several people who were at the center of events during all or part of the Carter and Reagan administrations. I wish to acknowledge the valuable insight that Omi Walden, Patty Bartlett, and Gary Moore provided me on the events and circumstances that shaped the program over those years. I also wish to thank my friend and former boss, Robert San Martin, for his long weekly staff meetings during which I had ample time to take the notes that formed an important record for one of the chapters.

When the drafts of chapters 5 and 6 were finally ready for outside review, Donna Fitzpatrick and Robert San Martin generously agreed to perform that task. They provided valuable comments, corrections, and verifications that, I hope, improved the accuracy of my recollections. Information regarding the Department of Energy reorganizations was furnished in part by J. Kenneth Schafer ("Organization Theory and Federal Agency Reorganizations: A Department of Energy Case Study," doctoral dissertation, George Mason University, 1994). Ken Schafer also provided access to file material and notes associated with his research. His assistance is greatly appreciated. *The Solar Energy Intelligence Report* provided an excellent memory jog for the endless stream of events that formed the solar energy program during the years covered in the series.

—Frederick H. Morse

Series Acknowledgments

Charles A. Bankston

As we complete the ten-volume series with the publication of this volume, I, as editor-in-chief for the series, would like to express my sincere thanks to all those who have contributed in large and small ways to the completion and success of the project.

The entire project is the result of the vision of Fred Morse, who conceived the series as a means of distilling and preserving the accomplishments of more than a decade of government solar energy research, development, and marketing activities for the benefit of future researchers and decision makers. His leadership, dedication, and determination guided the project through a myriad of bureaucratic, financial, and editorial obstacles. His enthusiasm, energy, and humor sustained me and many other contributors when it seemed that the project would surely succumb to one or another quagmire. When Fred left DOE in 1990, with the main elements of the series already complete or in progress, the project benefited from the continuing support of Bob Hassett.

My special and heartfelt thanks go the authors, editors, and reviewers who contributed their time, energy, and intellectual expertise to the project. It was indeed an honor and a privilege for me to work with the more that 200 leaders in the field of solar energy who helped create this legacy of solar energy knowledge. It has been especially gratifying to work with my friends and colleagues Don Beattie and Fred Morse on this volume.

I also wish to acknowledge the many contributors whose names do not appear at the head of volumes or chapters of the series, but whose efforts made my job and the jobs of the editors and writers possible, easier, or more fun. As my assistant for most of the project, Lynda McGovern-Orr kept track of the details and schedules of producing and reviewing outlines, extended outlines, drafts, and final manuscripts, helped organize editorial board meetings, produced progress reports, and generally helped solve problems and made everyone feel good about the project. Robert LeChevalier and Anne Townsend were of great value during the early stages of the project and helped to identify resources for authors and provided invaluable program memory. In the later stages of the project, the management and administrative details were handled by CBY Associates Inc.; I am especially grateful to Anne Harris, who helped with administration and production of several volumes, and to my partner

Sharon Yenney, who had the patience to put up with my seemingly endless involvement in the project and to cheerfully help out with whatever tasks needed to be done.

I cannot fail to acknowledge the tremendous support given the editors, authors, and myself by the publications group at SERI, including Paul Notari, Charles Berberich, editors Nancy Reese and Barbara Miller, and word processor Judy Hulstrom, and the project support of SERI director Harold Hubbard.

The author, editor, and project management contracts were administered by the Energy Technology Engineering Center, ETEC, a part of the Rocketdyne Division of North American Rockwell in Canoga Park, California. The efforts of ETEC managers Jack Roberts and Oscar Hillig to keep the project moving and resolve minor or major problems with financing and contracts were appreciated by all.

Finally, it has been a pleasure to work with Editor-in-Chief Larry Cohen and Melissa Vaughn at MIT Press. We are all grateful for Larry's confidence and patience when it seemed that some of the volumes would never be completed. Melissa skillfully guided several of the volumes through the MIT Press editing and production stages and helped make the series one of which we are all proud.

List of Acronyms

AEC	Atomic Energy Commission
AIA	American Institute of Architects
ANL	Argonne National Laboratory
AOO	Albuquerque Operations Office
ARI	Air-Conditioning and Refrigeration Institute
ARPA	Advanced Research Projects Agency
ASCS	Assistant Secretary for Conservation and Solar (Applications)
ASE	Alliance to Save Energy
ASHRAE	American Society of Heating, Refrigerating, and Air-Conditioning Engineers
BEPS	Building Energy Performance Standards
BLM	Bureau of Land Management
BNL	Brookhaven National Laboratory
BOB	Bureau of the Budget
CAFE	Corporate Average Fuel Economy
CCMS	Commission on the Challenges of Modern Society (NATO)
CEQ	(President's) Council on Environmental Quality
CNRS	National Center for Scientific Research (France)
COPEP	Committee on Public Engineering Policy
CORECT	Committee on Renewable Energy Commerce and Trade
CPC	compound parabolic concentrator
DHW	domestic hot water
DOC	U.S. Department of Commerce
DOD	U.S. Department of Defense
DOE	U.S. Department of Energy
DOE FO	DOE Field Office
DOI	U.S. Department of Interior
DPR	Domestic Policy Review
DSM	demand-side management
ECPA	Energy Conservation and Production Act

EER	Energy Efficiency Ratings
EIA	Energy Information Administration
EPA	Environmental Protection Agency
ECPA	Energy Conservation and Production Act
ERDA	Energy Research and Development Administration
ERSATZ	(synthetic solar radiation database)
ERTA	Energy Research and Technology Administration
FCST	Federal Council on Science and Technology
FEA	Federal Energy Administration
FEMP	Federal Energy Management Program
FEO	Federal Energy Office
FPC	Federal Power Commission
GAO	General Accounting Office
GATT	General Agreement on Tariffs and Trade
GNMA	Government National Mortgage Association ("Ginnie Mae")
GNP	gross national product
GSA	General Services Administration
HUD	U.S. Department of Housing and Urban Development
HVAC	heating, ventilation, and air-conditioning
ICC	Interstate Commerce Commission
ICPC	integrated compound parabolic concentrator
ICS	integrated collector and storage
IEA	International Energy Agency
IERS	Inexhaustible Energy Resources Study
IPH	Industrial Process Heat
IPTASE	Interagency Panel for the Terrestrial Applications of Solar Energy
IRP	integrated resource planning
IRRPOS	Interdisciplinary Research Relevant to Problems of Our Society (NSF)

List of Acronyms

ISCC	Interstate Solar Coordinating Committee
ITC	InterTechnology Corporation
JCAE	(Congressional) Joint Committee on Atomic Energy
JPL	Jet Propulsion Laboratory
LANL	Los Alamos National Laboratory
LASL	Los Alamos Scientific Laboratory
LBL	Lawrence Berkeley Laboratory
LERC	Lewis Research Center (NASA)
LLL	Lawrence Livermore Laboratory (became SNLL)
MASEC	Mid-America Solar Energy Complex
MHD	magnetohydrodynamics
MOPPS	Market-Oriented Program Planning Study
MOU	Memorandum of Understanding
MRI	Midwest Research Institute
MSFC	Marshall Space Flight Center (NASA)
NACA	National Advisory Council on Aeronautics (became NASA)
NAE	National Academy of Engineering
NAS	National Academy of Sciences
NASA	National Aeronautics and Space Administration
NATO	North Atlantic Treaty Organization
NBS	National Bureau of Standards (became NIST)
NEP	National Energy Plan
NESEC	Northeast Solar Energy Center
NIST	National Institute of Standards and Technology (was NBS)
NOAA	National Oceanic and Atmospheric Administration
NRC	Nuclear Regulatory Commission
NREL	National Renewable Energy Laboratory (was SERI)
NSF	National Science Foundation
NWS	National Weather Service
OCR	Office of Coal Research
OMB	Office of Management and Budget

OPEC	Organization of Petroleum Exporting Countries
OST	(President's) Office of Science and Technology
OTA	Office of Technology Assessment
OTEC	ocean thermal energy conversion
POA	Program Opportunity Announcement
POCE	Proof of Concept Experiment
PON	Program Opportunity Notice
PRDA	Program Research and Development Announcement
PURPA	Public Utilities Regulatory Policy Act
PV	photovoltaic
R&D	research and development
RANN	Research Applied to National Needs (NSF)
RCL	rating, certification, and labeling
RCS	Residential Conservation Service
RD&D	research, development, and demonstration
RFP	Request for Proposals
RIF	reduction in force
ROR	rate of return
RSEC	regional solar energy center
SAHP	solar-augmented heat pump
SAIC	Science Applications International Corporation
SCE	Southern California Edison
SDHW	solar domestic hot water
SEB	Source Evaluation Board
SEGBP	Solar Energy Government Buildings Program
SEGS	Solar Electric Generating System (Luz International)
SEIA	Solar Energy Industries Association
SEIDB	Solar Energy Information Data Bank
SEREF	Solar Energy Research and Education Foundation (SEIA)
SERI	Solar Energy Research Institute (became NREL)
SFBP	Solar in Federal Buildings Program

List of Acronyms

SIPO	Solar Institute Project Office
SNLA	Sandia National Laboratory Albuquerque (was Sandia Laboratories)
SNLL	Sandia National Laboratory Livermore (was LLL)
SOLMET	(solar radiation and meterological database)
SPS	solar satellite power system
SRCC	Solar Rating and Certification Corporation
SSEC	Southern Solar Energy Center
ST	Solar Thermal
SUN	(Western) Solar Utilization Network
TMY	typical meteorological year
TRW	Thompson, Ramo, Woolridge, Inc.
UNESCO	United Nations Educational, Scientific, and Cultural Organization
USDA	U.S. Department of Agriculture
USGS	U.S. Geological Survey
ZBB	zero-based budgeting
ZEG	zero energy growth

1 About this Volume

Charles A. Bankston and Donald A. Beattie

This ten-volume series was conceived in 1982 by Frederick H. Morse, then director of the Office of Solar Heat Technologies, Department of Energy (DOE). The purpose of this series is to provide a thorough technical assessment of the federal program in solar heat technologies. Leading experts from both the public and private sectors have documented, in detail, the most significant results and findings of the program with the overall goal of providing both a starting point for the new researcher and a reference tool for the experienced worker and manager. This volume presents a history and overview of the program from its inception at the National Science Foundation (NSF) in the early 1970s through 1988, the end of the Reagan administration. A cursory description of some of the events and changes that occurred in the Bush and Clinton administrations is included, but the program is not followed in detail beyond the second Reagan term. Some of the program's technical accomplishments completed after Reagan left office are mentioned here and in some of the other volumes of the series. Our budget charts, however, do track budgets through fiscal year 1995, the current year budget as this volume goes to the publisher.

Although the series as a whole is limited to solar heat technologies that convert sunlight to heat to provide energy for applications in the building, industrial, and power sectors, this volume briefly discusses all aspects of the federal solar energy research, development, and demonstration (RD&D) programs and helps place the solar heat technologies in the context of the much larger program.

Taking a broad, historical viewpoint, we have tried to provide the reader with an overall assessment of the accomplishments of and lessons learned from the program. We have discussed the events that shaped the program as well as the people and organizations that contributed to its successes and failures. We hope to have provided future decision makers who may pass this way again with a perspective on what does and does not work when the federal government undertakes large-scale energy RD&D programs and attempts to enter the field of commercialization, where success requires a major involvement of the private sector.

If the reader perceives a sense of struggle as the program unfolded, an "us versus them" tone, it is because that was the climate of the day. As this volume developed, three basic themes became evident without any prearranged discussion on the part of the authors. First, throughout much of the program's life, and certainly in the early and later years, many important decision makers in the executive branch, and some outside the government, strenuously opposed a strong federally sponsored solar program. Second, in spite of this opposition, an aggressive and successful program evolved and provided the country with concrete results. Third, even though the program expanded rapidly and solar energy systems of all types were quickly demonstrated, some prematurely, many solar enthusiasts were highly critical and felt the program should have moved much faster.

We will leave it to the reader to take sides and select which, if any, of these themes accurately reflects the program's history. As in almost all activities of this type, there is no black-and-white interpretation, only shades of gray. Undoubtedly, program management and supporters in and out of government could have made better decisions at many points along the way, although identifying these was possible only after the fact. The authors hope their insights will permit the next generation of managers and legislators to improve on the ongoing program or construct a better program when the nation needs it.

1.1 Setting the Scene

Government support of solar energy research has a long history. However, it was only in the early 1970s that this support began to mushroom into a major program. The authors of this volume were involved from the beginning in shaping and managing a rapidly growing program that was controlled, in the main, from Washington, D.C. Federal programs in solar energy RD&D evolved through a series of agencies and managers, driven by external events that included energy crises, administration changes, and close congressional oversight. These complex interactions, not unusual in government programs, had significant impacts on both program content and program results. The time line in figure 1.1 attempts, in one graphic overview, to help the reader place these events in historical perspective along with some of the global events that influenced U.S. energy policy. Most of the events shown in this figure will be discussed in varying degrees of detail in this volume. Figure 1.1(a) shows the major political and legislative events that shaped the solar energy program; figure 1.1(b) shows some of the world events and influential studies

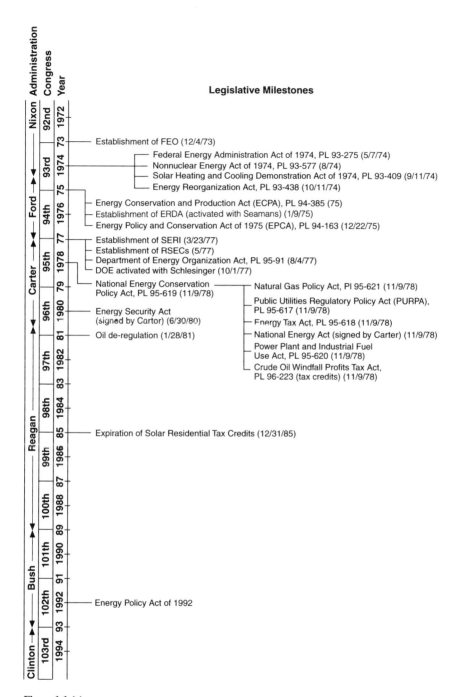

Figure 1.1 (a)
Time line for major political, legislative, and policy events that shaped the solar heat technologies programs of the 1970s and 1980s.

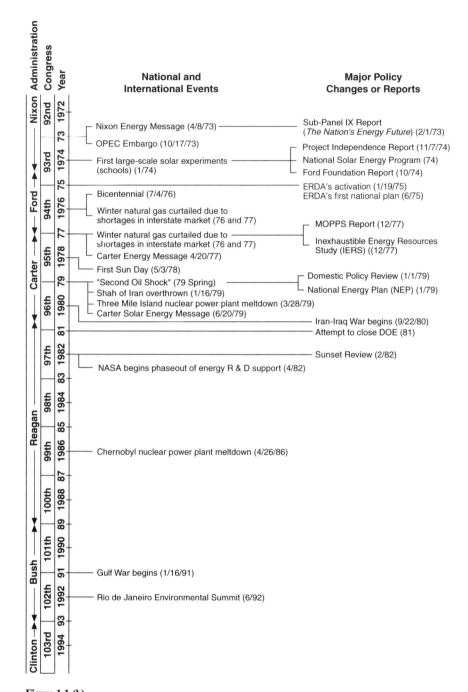

Figure 1.1 (b)
Time line for major studies and world events that shaped the solar heat technologies programs of the 1970s and 1980s.

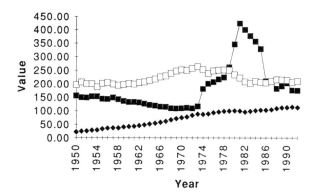

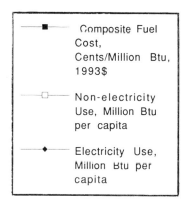

Figure 1.2
Composite energy cost, electricity and non-electricity use vs. time.

and reports of the period. But it was the escalating price Americans had to pay for energy from conventional sources and the importance of having a secure energy supply that drove the country to expand the RD&D programs on all forms of energy supply (including solar and other renewable energy resources), energy efficiency, and conservation. To illustrate this last point, we have included a graph of the composite energy cost and per capita energy use in figure 1.2 to show what was happening in the marketplace.

We suggest that the reader return to these figures from time to time while reading subsequent chapters. They will help in understanding what was going on outside the solar energy program and keep the chronology of the program straight when our text, for the purposes of continuity, departs from a strictly chronological narration of events to follow a particular program or institution over a period of time.

1.2 The Budgets

Because our discussion of the solar energy program is intrinsically tied to the federal budget, the reader will find budget figures in every chapter. Although the precise budget numbers are not important, they often indicate the popularity of the program with a particular administration, congressional committee, or bureaucrat. Further, the reader should understand that at any given time a government manager is usually dealing with three budgets simultaneously: (1) the appropriated funds being allocated in the current fiscal year, (2) the budget before Congress for the next fiscal year, and (3) the budget being prepared for submission to the Office of Management and Budget (OMB) for two fiscal years down the road.

As useful as budget figures are in conveying the status and popularity of programs, they can also be confusing for the reader (or the researcher). An agency budget begins as low-level staff working documents, which are reviewed at senior agency levels before being unified into an agency request, which in turn is negotiated and approved by the incumbent administration (OMB) before being submitted to Congress. Once in Congress, the administration request is usually modified by different subcommittees of both houses, the subcommittee versions must be reconciled and voted on by both houses, and the Senate and House versions must be negotiated in committee and approved (appropriated) with final votes in both houses. Only then does the budget go to the president for signature.

Even then, an approved budget may not be what is actually available to the requesting agency. The administration may withhold or rescind certain parts of the appropriation, or the agency itself may redistribute some of the funds to different programs. The upshot of all of this is that the records of a program may contain references to many different figures as "the budget" for a particular program in a particular year, depending on the stage of the budget cycle in which the document was written. To further complicate the accounting, a particular program may be included in the budget for one organization one year but be shifted to a different organization the next. Some programs may receive funds from more than one organization or even more than one agency.

Although all these budget complications befell the solar energy program at some time during its evolution, the reader need not be concerned with the absolute accuracy or consistency of the budget figures cited in this text or in other volumes of this series. Wherever possible, we have tried either to specify the phase of the budget cycle to which a given figure pertains, or to use the final budget made available to the program. In subsequent chapters, we give budget figures in current year dollars. However, part of the solar era was also an era of high inflation, and the value of the dollar fell by about two-thirds between 1973 and 1995. To allow today's reader to better appreciate the size of the overall program and the spending that took place, we have included the Gross National Product Implicit Price Deflators used to convert from current to constant dollars. These data are shown in table 1.1, along with the total federal solar energy program budgets for FY 1974 through FY 1995 (estimated) in both current and constant (1993) dollars, displayed graphically in figure 1.3. Because this series deals only with the solar heat technologies, we also show the fraction of the total solar budget that was allocated to these programs alone. Figure 1.4 shows the annual budgets for the Active, Passive, and Solar Thermal programs in current year dollars as compiled from DOE sources and the authors' personal files. Note that in 1985 the Active and Passive Programs were combined to form the Solar Buildings Program. The major fraction of the funding difference between the total solar budget and the solar heat technologies budgets was for the photovoltaic program, although the total also includes funding for smaller programs such as wind energy, biomass, and ocean thermal energy conversion (OTEC).

If the reader sometimes wonders why the solar energy program appeared to be in such a state of flux, a quick return to figures 1.3 and 1.4 will probably clarify the central problem.

1.3 The Organizations

Today, the federal solar energy RD&D program is largely the domain of the Department of Energy (the EPA has some solar programs as well), but it was not always the case that a single agency was in charge. In the early years many agencies were involved; some were formed and dissolved or entered and left the program within the period covered by this volume. Much of chapters 3 and 4 deals with the interagency ebb and flow of programmatic responsibilities. The reader may find this somewhat

Table 1.1
History of the total solar or renewable energy program budget in current and 1993 dollars

Fiscal year	GNP deflator (EIA data)	Federal budget (current year $)	Federal budget (1993 $)	Total cumulative (current year $)	Total cumulative (1993 $)
1974[a]	44.9	20.6	57.0	20.6	57.0
1975	49.2	48.0	121.2	68.6	178.2
1976	52.3	158.7	376.9	227.3	555.0
1977[b]	55.9	282.2	627.0	509.5	1,182.0
1978	60.3	408.8	842.0	918.3	2,024.0
1979	65.5	484.3	918.3	1,402.6	2,942.4
1980	71.7	548.4	949.9	1,951.0	3,892.3
1981	78.9	548.9	864.0	2,499.9	4,756.4
1982	83.8	221.3	328.0	2,721.2	5,084.3
1983	87.2	197.0	280.6	2,918.2	5,364.9
1984	91.0	176.5	240.9	3,094.7	5,605.8
1985	94.4	169.8	223.4	3,264.5	5,829.2
1986	96.9	143.2	183.5	3,407.7	6,012.8
1987	100.0	122.5	152.1	3,530.2	6,164.9
1988	103.9	96.9	115.8	3,627.1	6,280.7
1989	108.5	91.5	104.7	3,718.6	6,385.5
1990	113.3	92.4	101.3	3,811.0	6,486.8
1991	117.7	129.5	136.7	3,940.5	6,623.4
1992	121.1	175.5	180.0	4,116.0	6,803.4
1993	124.2	181.6	181.6	4,297.6	6,985.0
1994	127.2	238.0	232.4	4,535.6	7,217.4
1995	130.2	289.0	275.7	4,824.6	7,493.1

Source: DOE Energy Information Administration.
Notes: Although the program name changed in 1981 from "solar" to "renewable" energy, the program contained the same activities.
a. For solar budgets prior to FY 1974, see chapter 3.
b. In FY 1977 the federal government changed its fiscal year start and end dates. To accommodate this change, FY 1977 includes five quarters.

About this Volume 9

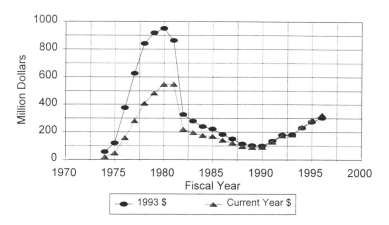

Figure 1.3
History of federal solar or renewable energy budget in current year and 1993 dollars. Source: DOE Energy Information Administration.

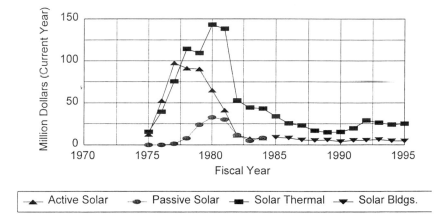

Figure 1.4
Breakdown of federal solar budget by solar heat technologies. Sources: DOE EIA, and personal files.

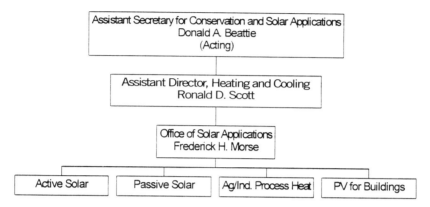

Figure 1.5
Solar heat technologies program organization in 1978.

confusing without a road map. Figure 4.1 displays the evolution of agencies' involvement in the solar energy program and their major program responsibilities.

By the beginning of 1978, the DOE had been operating for four months and had taken over most of the solar work previously spread among many agencies. But the changes did not stop. Each successive administration found it necessary to reorganize the part of the DOE that dealt with solar and other renewable sources of energy. Sometimes there were major reorganizations within an administration's tenure. These reorganizations are a dominant theme of chapters 5 and 6, where the influences of the organizations and the individuals who led them are discussed. Figures 1.5–1.8 show four simplified organizational charts for the solar energy program within the DOE from the Carter to the Bush administrations. Chapters 5 and 6 place individuals in some of the positions shown in these charts. (Perhaps here is a good place to apologize to the reader for the frequent use of acronyms, which are endemic to all government programs; to make partial amends, we have included in the front matter an acronym list that should prove useful as you proceed through this and the other volumes of the series.)

1.4 The RD&D and Commercialization Program Support Structure

While the solar energy program was managed by the agencies in Washington, D.C., most of the work was in fact done by private sector

About this Volume 11

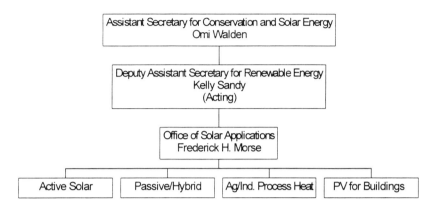

Figure 1.6
Solar heat technologies program organization in 1979.

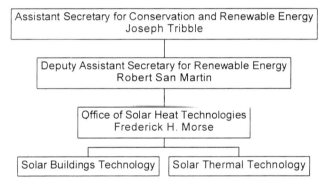

Figure 1.7
Solar heat technologies program organization in 1981.

companies, by laboratories attached to the agencies, or by universities receiving federal grants and contracts. Work continued to be performed in this manner as management was transferred from one agency to another and as new agencies and laboratories were formed. Figure 1.6 depicts the government program support structure as it was near the end of the 1970s. At that time the program included most of the early participants from NASA and AEC programs as well as the newly formed DOE institutions. In subsequent years, as program budgets declined and many of the field organizations dropped out of the program, their functions were consolidated at the Solar Energy Research Institute (SERI);

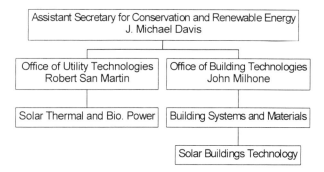

Figure 1.8
Solar heat technologies program organization in 1990.

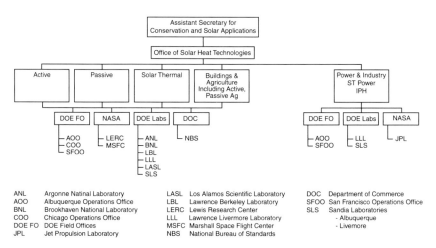

Figure 1.9
Solar energy program field support structure.

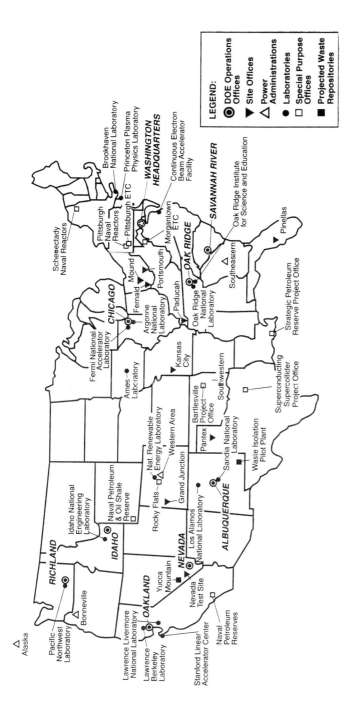

Figure 1.10
Location of DOE laboratories and field offices.

later the National Renewable Energy Laboratory or NREL) and the few other national laboratories still involved (figure 1.10 shows the location of the laboratories and operations offices listed in figure 1.9).

1.5 The Results

The other volumes of this series, with the exception of volume 10, concentrate on the results and accomplishments of the solar heating technology RD&D programs supported by the federal government. They contain details, often in the form of equations, graphs, tables, diagrams, or photos, that will be interesting primarily to researchers or practitioners of solar heating technologies. In this volume we provide a more general view of the most important findings and accomplishments. Chapter 7 is a relatively nontechnical, personal assessment by the editor-in-chief of the most significant R&D results covered in all nine volumes and some of the results that have come to light since those volumes were written. While chapter 7 covers the same topics as the other series volumes, it is not a summary and does not make specific reference to the other volumes. It is primarily for the less technical reader who would like to know something about what came out of this undertaking without reading the thousands of pages that constitute the entire series. Chapter 8 summarizes the authors' views on the lessons learned, both programmatic and political; it is our hope that technical readers will find these observations both accurate and relevant, and will be stimulated to delve into the other volumes for more specific information.

2 About this Series

Charles A. Bankston

Most fields of research in the United States have been influenced, if not dominated, by government-sponsored programs. Nuclear energy is certainly an example where government programs shaped the course of development of an energy technology. Exploitation of the country's vast coal resources was made possible by early research and surveys conducted by the U.S. Geological Survey (USGS) and the Bureau of Land Management (BLM). In addition to government-conducted research, government actions such as tax incentives have provided early stimulus to the utilization of energy resources such as oil and gas. In the field of solar energy, in its many forms, government programs have thus far dominated its research and subsequent development.

The development of industries for the conversion and utilization of most of our natural resources, whether aided by government or not, has been fairly smooth and continuous, if not always rapid. Coal, oil, natural gas, and uranium have progressed from discovery to widespread use in a matter of decades and have been in continuous use, some for centuries. Although there have been plateaus, none of these resources have vanished from use, or even declined in absolute terms. Solar energy technologies, on the other hand, have experienced several periods of acceptance and growth, only to fall back into disuse and relative obscurity when more easily exploited resources were discovered or recovered. An erratic growth pattern makes it difficult for any technology to become economically accepted. In a start-and-stop development cycle, technical expertise, the industrial base, and the supporting infrastructure tend to fade away and must then be rebuilt if the technology reemerges. Unlike previous episodes of solar development, which were highly localized and economically driven, the development of solar heat technologies in the 1970s and early 1980s was more rapid and in some ways less orderly than the development of other energy technologies. It occurred in a near-crisis atmosphere in a diffuse geographical setting. There was little concentration of expertise and even less concentration of production capacity and infrastructure. Although much of the research and development was funded by government, with its well-known proclivity for reporting results, and followed well-conceived and well-implemented plans, the goals that had been set were generally not reached, and the successes the programs enjoyed were temporary and partial. For the most part, the advances achieved in the laboratories of science and industry did not find their way into the marketplace, and sometimes not even into the technical literature.

In a free and competitive economy, little can be done to preserve the production capacity and infrastructure on which a technology depends if it is not accepted into the marketplace, but at least a part of the knowledge base can be mothballed in the archival literature. The U.S. federal government spent more than 1.7 billion dollars (3 billion 1993 dollars) of the taxpayers' money to advance the science and technology of solar heating and more than 5 billion current dollars (7.8 billion 1993 dollars) for all solar technologies. This series is an effort to retain as much as possible of the knowledge base those billions of dollars purchased. Although it involved many authors and editors over a period almost as long as the solar era it covers, because the compensation of authors was minimal, the dollar cost was small in comparison to the cost of reconstructing the knowledge base or of retrieving it from the bits and shards of the trash heap of terminated programs and abandoned projects.

2.1 Background

In the months following the election of 1980, the implications for the future of energy policy in the United States began to emerge. The Heritage Foundation, a conservative think tank, assembled teams of "advisors" to help shape President Reagan's nominations and policies. The chairman of the transition team and many contributors to the energy panel came from the oil and gas industry. Besides their conservative bent for less government involvement in domestic affairs, they did not support the commercialization of renewable energy options. The recommendation from Heritage was that the entire Department of Energy, along with the Department of Education, be abolished. Nuclear weapons programs within DOE would be moved to the Department of Defense. A few of the other departmental activities might be assumed by the Department of Commerce or the Department of the Interior, but the clear message was that the administration would no longer support alternative forms of energy. Because the abolition of the Department of Energy met with strong congressional opposition, a dentist (who was also an ex-governor and strongly pro-nuclear) was appointed Secretary of Energy to emphasize the low esteem in which the department was held by the administration.

Each day of the new administration weakened the department. Budgets from OMB recommended immediate closeout of most if not all of the department's renewable energy and conservation programs. Staff allocations were drastically cut, and the reduction in force (RIF) song echoed through the halls of the Forrestal Building. Some bureaucrats scrambled to find jobs in politically correct programs, some stood and fought for their programs and budgets, and a few worried about what would become of the advances they had labored long and hard to achieve. The solar programs had grown so fast and moved ahead so rapidly that little thought had been given to assessing and documenting their progress. Few programs had been completed, and the policies intended to encourage adoption of solar technologies (commercialization) had just been put in place. Nevertheless, the mood of the administration was to ax programs first and ask questions later.

Director of the Solar Heat Technologies Office of DOE Frederick H. Morse had the vision and the courage in early 1982 to initiate a project of systematic assessment and documentation of the research and development accomplishments of the preceding decade of government solar thermal programs. Director Morse intended the effort as an attempt neither to glorify the past programs nor to justify their expenditures. Rather, as he put it, "it was an obligation to report the important program achievements and to assess their impact on technology and utilization of solar thermal energy" for the benefit of future researchers and planners.

Although there was strong commitment from Morse and a few others within the department, the "Solar Thermal Energy Conversion Program Documentation Project," as it was then called, frequently met with opposition from above, below, and afar. Some newly appointed functionaries at DOE were suspicious of the motives of the project. They worried that disappointed researchers and bureaucrats would use the project to ridicule the administration or to try to build public or congressional support for programs that the administration wanted canceled. Some program managers within the Solar Heat Technologies Office considered the project, or any new projects, as unwelcome competition for the reduced budgets that remained. Some laboratories and field offices worried that the assessment might reflect badly on their performance or management. In fact, the project did all these things, but not to an extent that would cause any problems. Time took care of that.

2.2 Project Objectives

Although the scope of its coverage changed as time went on, the project's objectives and basic approach did not. These were clearly stated at the outset in the original "Objective Statement" of September 16, 1982, reproduced below.

OBJECTIVE

SOLAR THERMAL ENERGY CONVERSION PROGRAM DOCUMENTATION PROJECT

The Solar Thermal Energy Conversion Program Documentation Project has been initiated by the Office of Solar Heat Technologies, U.S. Department of Energy (DOE), in response to a recognized need to identify and concentrate in a single, readily accessible group of integrated publications, the most significant and useful results for Federally supported programs. It will document the advances in the state-of-the-art resulting directly or indirectly from a decade of substantial Government support of solar thermal energy conversion activities. The objective of the project is to produce a high-quality series of manuscripts, organized, written and edited by leading authorities in the solar energy community, which can be published by the private sector as a landmark in the growth and development of solar energy.

The Project's purposes are manifold. It will provide an objective and comprehensive assessment of major accomplishments of the various programs and significant lessons learned through research and field experience. It will provide a compact but a comprehensive compendium of the most important research findings and results. It will provide an objective identification of the limitations of our knowledge of the theory and applications of solar thermal energy conversion for all applications and of the opportunities for research to reduce these limitations. Finally, it will provide a technically sound digest of research results that can be readily translated into practical rules and guidelines for non-technical specialists and the lay community.

The Project approach is patterned after the private sector. The contributors, both authors and editors, will be selected from leaders of the solar community on the basis of their knowledge, ability, and objectivity. Contributions by authors will represent their own analyses of the subjects reviewed and will be so identified and acknowledged. The organizational affiliation of the authors and editors will be acknowledged, but no element of the publication will be institutionalized and the authors and editors, not their institutions, will be responsible for the quality and timeliness of their contributions. The authors, or their institutions if they are affiliated with Government agencies or laboratories, will be paid a small stipend, but the real motivation is the opportunity to be identified with a prestigious publication, to work with editors who are acknowledged leaders in the solar community, to learn and profit from an in-depth review of their field of specialization, and to have the opportunity to influence the direction of future research, development and application of solar thermal energy conversion.

The target audiences will include researchers in academia, industry and Government, those who plan or sponsor private or public supported research, and those involved in converting research into practical services or products. The organization and structure of the manuscripts will be topical, with sections devoted both to generic subjects such as the solar resources and to specific applications such as solar water heating. The content will be technical with important results presented in concise narrative, tabular, graphical or mathematical form. Only the history and background necessary to put the results in perspective and highlight their significance will be included. All material will be supported with verified references to the most readily available technical literature. The reference base will be programmatic and will include all relevant Federally supported work including DOE and other agencies' programs. Although the emphasis will be on accomplishments of the Government programs, work accomplished in the U.S. private sector or internationally will be included to the extent necessary to define the state-of-the-art and to place the results in perspective.

Each major subject will be covered by one or more authors working under the direction and guidance of a manager (editor) who will be responsible for attaining the proper program coverage, balance, depth of treatment of the subject, and technical quality. The Project will be directed by a Project Director (Editor-in-Chief), who will be responsible for the overall content, quality and schedule. The Technical Information Division of SERI will provide editorial support to assure consistent editorial style, usage, format, nomenclature and graphics, and will represent the project in negotiations with private publishers.

The complete project will consist of manuscripts for six volumes as shown in TABLE 1. At present, editors and target completion dates have been selected for Volumes II, III and IV only. The Project Director (Editor-in-Chief) is Dr. Charles A. Bankston, who may be contacted at (202) 338-2725 for more information.

Dr. Frederick H. Morse,
Originally Signed
September 16, 1982

2.3 Approach

The documentation project was managed by the editor-in-chief, who operated under an independent contract. The editor-in-chief was assisted by an editorial board that included some of the leading authorities in the field, as well as representatives of the three divisions of the Office of Solar Heat Technologies, the office's director (Frederick Morse), and a self-proclaimed guardian of the administration's dogma (Jay E. Holmes). The board's functions were to help the editor-in-chief in structuring and staffing the project, to provide guidance on the topics and scope of the volumes, to review and approve the volume outlines and authors, and to help form editorial policies for the series. It met frequently and participated in a computer conferencing electronic network during the formative stages of the program; most of its members became volume editors.

As stated in the Objective Statement, the authors and editors of the topical volumes were to be selected for their knowledge, ability, and objectivity. They were to research and present their own assessment of the work in their subject area without influence from their institutions. This was accomplished by issuing individual contracts for authors and editors. Although the remuneration was nominal, the contract bound the parties to the thorough and complete process; it was a bond that had considerable elasticity.

The selection of authors was the responsibility of the volume editor, but the editorial board offered suggestions and recommendations. DOE approved the authors and the preliminary outlines of their chapters. No subsequent DOE review or approval of the material was required, although DOE personnel occasionally participated as technical reviewers for specific chapters. This was an important decision made by Director Morse to ensure that the work would not be biased by program managers who might be interested more in justifying their current program directions than in an objective assessment of accomplishments.

The volume editors were responsible for recruiting authors and for the technical content and editing of all materials. They reviewed and approved preliminary and extended outlines of each chapter before the authors began writing. A first draft was reviewed by the volume editor, and these recommendations were incorporated into a revised draft, which was then sent to at least two independent reviewers. The suggestions and criticism of the reviewers were then coordinated by the volume editor, who had final responsibility for deciding what revisions were needed. In almost all chapters, this review and revision process was friendly and constructive, but the editors were not required to, and usually did not, reveal the identity of the reviewers to the authors. This rather formal and very thorough review process was a major strength of the project, but it also added a great deal of time and managerial effort.

Because the project was started in the very early days of the personal computer era, and because the intent of the project was that the contributions be personal, not institutional, authors were invited to submit their work in any form, from handwritten text to dictation tapes, to the volume editor. The editor would then submit the chapters individually or collectively to the word processing group at SERI for computer entry and line editing. This step was meant to free the authors from the need to type manuscripts and to ensure a consistent style and format for the entire series. It was also intended to simplify the final publication of the volumes by having the entire volume available in electronic form for the typesetter. As the projects progressed, this process became less important and was eventually abandoned. SERI (or NREL, as it was later known) did not participate at all in the preparation of the last three volumes produced (volumes 4, 10, and 1).

To fulfill this objective of publication in the private sector, the editor-in-chief, assisted by SERI's technical publication staff, made initial contact with a wide range of professional societies and publishers. In 1983 SERI formally contacted more than 300 publishers regarding the project and received expressions of interest from about a dozen.

In 1984 meetings were held with representatives of the professional societies to inform them about the project and to assess their interest in and support for the publication. Following these meetings, SERI arranged a briefing in New York City on December 7, 1984, at which the publishers were able to discuss the project with the editor-in-chief and the director. Proposals were then requested, and in the summer of 1985 SERI selected the MIT Press on the basis of the prestige and independence of its technical list and its strong marketing plan. Larry Cohen, then science editor and now editor-in-chief of the press, was designated to oversee publication of the series.

At Larry Cohen's first meeting with the editorial board on March 17, 1986, the six volumes mentioned in the project objective statement above were rearranged into ten volumes, and two additional volumes were added. The two new volumes, dealing with the solar thermal power and industry technologies, were listed in the early series pages as volumes 10 and 11, but they were eventually dropped from the series when it became obvious that there was little support for their preparation. The volume dealing with the implementation of the technologies was moved up from volume 12 to 10 in the final listing, which can be found on the series page in this volume.

It was also decided at the 1986 meeting that the volumes would be published as completed rather than in the logical order in which they are numbered. The first volume to be published was thus volume 3, *Economic Analysis of Solar Thermal Energy Systems*, in 1988. Volumes 2 and 8 were published in 1989, volume 9 in 1990, volume 5 in 1991, volume 7 in 1992, and volume 6 in 1993; volumes 4 and 10 will both appear in 1996. One would like to think that the extended gestation period of the series, which none of us foresaw when we first mapped it out, has allowed us to build in a broader perspective on the technologies we have tried to cover. But present and future readers and researchers will have to be the judges of that.

3 The Early Years of Federal Solar Energy Programs

Donald A. Beattie

Federal support of solar energy research dates back to at least the 1940s, although private sector funding was the primary source that sustained early experimentation. Federal funding for solar research was small until a single event, the oil embargo initiated by the Organization of Petroleum Exporting Countries (OPEC) in October 1973, galvanized the government into concerted action across the energy spectrum. Newspaper and TV pictures of children in their coats, huddled in heatless classrooms, and long gas lines created an outcry to "solve the problem" and, in some quarters, a call for "energy independence." As a result, beginning in 1974, there was a dramatic increase in federal funds devoted to solar energy research and development.

Succeeding chapters will reconstruct the events that led to progress in solar energy RD&D and, where possible, provide some insight into the backstage maneuvering that always accompanies major government initiatives. At some points the discussion will apply to the entire solar energy program (including wind, biomass, photovoltaic, and other such systems), and at other points only to solar heat technologies. The authors will clarify these differences where appropriate. The remaining volumes of this series contain detailed descriptions of each of the solar heat technologies as well as discussions of their economics and related topics. Here I would like to start by outlining the foundations upon which a program was built that eventually grew to a yearly expenditure in FY 1980 of almost $550 million (including over $400 million for solar heat applications).

Attempts to use sunlight as a source of energy for processes ranging from water distillation, to home heating, to various types of steam engines go back hundreds, and for some applications even thousands, of years. Many of the early experiments were carried out by dedicated researchers operating with minimal resources. Early projects were carried out in many countries, including the U.S.S.R., Egypt, and France as well as the United States.[1] In the United States alone during the 1930s and early 1940s, over 100,000 solar water heaters were installed, primarily in Florida and southern California, while many simple, evaporative space cooling systems appeared in the Southwest. In the 1950s the U.S. Department of the Interior, Office of Saline Water, funded a number of experimenters for solar water distillation projects of various designs.[2] By the early 1970s, although advanced, commercially available solar water heating systems were being installed in many countries, most notably Israel, Japan, and Australia, such systems had fallen out of favor in the United States. Space heating, using various types of solar energy systems, had also advanced, but not to the point of widespread acceptance.

3.1 Initial Organized Activities

In the United States, solar energy research activities were sporadic at best until the 1940s, when a small group at the Massachusetts Institute of Technology (MIT) under the direction of Hoyt C. Hottel, who chaired the Solar Energy Coordinating Committee to oversee research among the various MIT departments, undertook a comprehensive research program. With financing provided primarily by the Cabot Fund, a series of houses was built over the next decade to test various approaches to the utilization of solar energy (figure 3.1 is an example of a typical house). This solar house program provided a training site for a number of solar researchers, who went on to productive careers at other institutions.

Figure 3.1
MIT Solar House I, November 1939. World Wide Photos. Courtesy of The MIT Museum.

As a result of the work at MIT, a space heating symposium was held in Cambridge, Massachusetts, in 1950. Follow-up symposia, seminars, and conferences were held throughout the 1950s and early 1960s at various locations in the United States and around the world as more researchers sought forums to discuss their work. A solar energy symposium at the University of Wisconsin in 1953, sponsored by the National Science Foundation, was followed by the 1955 World Symposium on Applied Solar Energy and the 1961 Conference on New Energy Sources, both sponsored by the United Nations, which examined the prospects for solar, wind, and geothermal energy. Growing interest in the practical applications of solar energy in many countries led solar scientists from around the world to form the Association for Applied Solar Energy in 1955, which eventually, in 1971, gave birth to the International Solar Energy Society, with sections in 10 countries and a membership of 4500. Thus by the early 1970s, there was a dedicated, exprienced cadre of scientists and engineers actively engaged in research on which the soon-to-be expanded federal efforts could draw.

In 1952 the President's Materials Policy Commission predicted that the nation's energy demand would double over the next quarter century.[3] The commission concluded that while conventional sources such as coal, oil, gas, and hydropower could fill this demand, there would be intense competition for these finite energy sources from both the developed and developing countries. Two new energy sources, atomic and solar energy, stood on the horizon as tremendous potential contributors. The commission suggested that the most important step the government could take was to develop a comprehensive energy policy that brought all the energy programs under one roof.

Although the comprehensive policy recommended by the commission was not forthcoming, and no federal programs of any magnitude developed for research in solar energy, the NSF did provide a small amount of funding to individual investigators over the next two decades to carry out research programs. These funds were augmented by foundations and other private sector sources and kept a small number of researchers busy in their laboratories.

By 1970, approximately twenty experimental solar-heated structures, using various combinations of collector types, heat storage techniques, heat transfer media, and backup energy sources, had been designed, built, and were operating in the United States. Most of the structures served primarily as laboratories for data collection, although some doubled as residences. In addition to the MIT group discussed earlier, George Löf at Colorado State University, Eric Farber at the University of Florida, Maria Telkes and Karl Boer at the University of Delaware, Harold Lorsch at the University of Pennsylvania, and John Duffie at the University of Wisconsin, to name but a few, were active researchers, all of whom were eventually supported by the NSF programs discussed below. The technology they employed typically used water or air collectors, insulated water storage tanks or rock bins for heat storage, and various types of pumps and controls. A few experiments had been conducted with solar absorption air-conditioning as well as other concepts such as dehumidifying agents with solar regeneration. The economics of these early systems were poorly understood, but indications were that mass-produced, more efficient solar collectors, along with tighter system design and integration, would permit solar energy to be cost-competitive in certain applications, especially if life-cycle costing were adopted.

3.2 NSF's Early Support

In 1968, eighteen years after the creation of NSF, Congress authorized the foundation to begin a program to focus on specific national problems. The next year, acting under its new mandate, NSF established Interdisciplinary Research Relevant to Problems of Our Society (IRRPOS), a small program with a modest budget of a few million dollars whose wide-ranging research included a number of projects related to energy and the environment. In a break with NSF tradition, the IRRPOS program instituted a new criterion for judging the value of proposals submitted by research institutions by requiring that they include a participant—in the private sector or in state or local government—who would act as a potential implementer if a useful end product of the research were achieved. For a project to receive NSF support, it was also expected that the end-use partner would also provide financial or in-kind support. This new way of doing business distinguished NSF's approach to applied research from its more traditional approach to basic research, where such associations were neither required nor expected.

In 1971, with the IRRPOS budget growing to $13 million, NSF created a new Directorate for Research Applications, and within this new directorate the IRRPOS program was renamed Research Applied to National

Needs (RANN). Alfred J. Eggers, Jr., a former NASA senior official, was appointed the first Assistant Director for Research Applications.

These activities coincided with major new national concerns. By the early 1970s, the energy problem forecast by the 1952 President's Materials Policy Commission and others became a reality, and sooner than anticipated. Domestic oil production peaked in the United States in 1970, and the country, which up to World War II had been a net oil exporter, was importing over 30 percent of its petroleum needs by the early 1970s. In his energy message of June 4, 1971, President Nixon called for a program to ensure that adequate supplies of "clean" energy would be available in the years ahead.[4] Under the direction of the President's Office of Science and Technology (OST), the Federal Council on Science and Technology (FCST), Committee on Energy R&D Goals established sixteen panels to study the complete spectrum of energy sources, including controlled fusion and solar energy. Although the objective of the panels was to recommend research and development goals for all energy technologies, for reasons not completely clear, the only panel report that was published dealt with solar energy.

The FCST Solar Energy Panel, established in January 1972 and jointly chaired by NSF and NASA, recommended a fifteen-year program that would cost in excess of $3.5 billion.[5] Participating on the Solar Energy Panel were forty-four "experts" from government, universities, and the private sector. Most of the "experts" came from outside the government (only three federal agencies—NSF, NASA, and DOD—were represented). Reflecting the panel's makeup, the recommended program included funds to be spent by the federal government and the private sector. In its conclusions and recommendations, the panel stated that "a substantial development program can achieve the necessary technical and economic objectives by the year 2020. By that date solar energy could economically provide up to (1) 35% of the total building heating and cooling load, (2) 30% of the nation's gaseous fuel, (3) 10% of the liquid fuel, and (4) 20% of the electric energy requirements." The panel also predicted that if the recommended R&D program were successful, "building heating could reach public use within five years, building cooling in six to ten years." With a draft of the Solar Energy Panel's recommendations in hand, NSF/RANN sent OMB its budget request for the first comprehensive terrestrial solar energy research program in October 1972, two months before the formal release of the FCST Solar Energy

Panel report. After a lengthy internal debate at NSF, the following three objectives were developed for the solar R&D program:

1. To provide the research and technology base required for the economic, terrestrial application of solar energy, and to foster the implementation of practical systems to the status required for commercial utilization;

2. To develop at the earliest feasible time the potential of solar energy applications as large-scale alternative energy sources;

3. To provide a firm technical, environmental, social, and economic basis for evaluating the role of solar energy utilization in U.S. energy planning.

Even at this early stage, it was recognized that bringing a new energy technology to the marketplace would require much more than technology R&D; many infrastructure and institutional restraints also would have to be overcome. These objectives, while appropriate for an applied research program, would later engender friction within NSF and among its academically based research clientele whose traditions favored rigid support of basic research.

Although greatly reduced in scope from the program called for by the Solar Energy Panel, the request to OMB outlined an R&D program that would cost $196 million from FY 1974 through FY 1978. Of this total, some $27.6 million was requested for solar heat technologies. Compared to ongoing major R&D programs for other energy technologies, such as atomic energy and coal, the modest NSF plan called for tripling FY 1973 expenditures, to more than $12 million in FY 1974. As shown in the projected funding patterns submitted with the plan, it was NSF's clear intent that this five-year period would represent just the beginning of a major effort to develop the many potential uses of solar energy.

In 1972 and early 1973 this proposed program went essentially unnoticed. Most of the many solar and environmental interest groups that today constitute a strong lobby for solar energy, did not exist at that time. Somewhat surprisingly, OMB accepted the proposed program, and steps were taken by NSF to implement it.

Just as the FCST Solar Energy Panel was developing its recommendations, two other energy reviews were undertaken, one by the National Academy of Engineering (NAE) and the second by the House Committee on Science and Astronautics, Task Force on Energy.[6] The House task force reported in December 1972 that "because of its continuous and virtually inexhaustible nature, solar energy R&D should receive greatly increased funding. Near-term applications of solar power for household uses seem likely, and central station terrestrial solar power and satellite solar power are attractive long-term possibilities." The NAE study was conducted by the Committee on Public Engineering Policy (COPEP) at the request of NSF/RANN to review RANN research priorities; its much-delayed final report, *Priorities for Research Applicable to National Needs*,[7] included a statement of strong support for continued solar energy R&D: "RANN support for applied research to improve prospects for the practical utilization of solar energy seems particularly attractive." Thus the stage was set for widespread support of solar energy research in response to the dramatic events of late 1973.

Although a small amount of funding in the IRRPOS and RANN budgets was earmarked for solar energy, as one can easily surmise from the less than generous totals shown in table 3.1, few projects could be supported, and individual grants were usually very small.

3.3 NSF Becomes the Lead Agency for the Solar Energy Program

As word of a forthcoming, expanded program for solar energy development spread, interest blossomed in the research community. Unsolicited proposals began to pour into RANN; it became very clear that new ideas abounded and that the prospects for advancing solar technologies based on new developments in many scientific and engineering fields were good.

Table 3.1
Solar energy technology funding ($ millions)

Technology area	FY 1971	FY 1972	FY 1973
Heating and Cooling	0.54	0.1	0.4
Solar thermal	0.06	0.55	1.43
Bioconversion	0.6	0.35	0.65
Photovoltaics	—	0.33	0.79
Wind energy	—	—	0.2
Ocean thermal	—	0.14	0.23
Workshops, etc.	—	0.19	0.25
Totals	1.2	1.66	3.96

The RANN staff administering the program was small, augmented by several university researchers brought in on detail for one-year assignments. Lloyd O. Herwig, Harold Horowitz, and Frederick H. Morse, the last on loan from the University of Maryland, had the responsibility to plan and oversee the NSF Solar Heating and Cooling programs. A staff goal was to distribute the small amount of funding to as many researchers as possible; by FY 1973, over fifty individual projects covering all solar technologies were underway. This was a somewhat remarkable achievement for, in spite of RANN's new charter, its grant proposals still went through a traditional NSF peer review process. To reduce the turnaround time, after a proposal had been reviewed by five or more impartial experts in the field, the cognizant NSF manager would make a recommendation to the NSF Grant Review Board, which in turn would say yea or nay. This process, from receipt of proposal to final approval, typically took six months or more (provided you could get your busy peer reviewers, working pro bono, to answer their mail; most did). Then, and only then, could you start final grant paperwork. In addition to accepting unsolicited proposals, RANN instituted a "program solicitation," modeled after a similar NASA process, to advertise to the research community opportunities to submit proposals in new areas of interest. In this same time period, reflecting the concern with moving ahead rapidly along all R&D fronts, NSF's contracting authority was expanded to permit RANN to contract directly with industry.

As the nation became more aware of U.S. dependence on foreign supplies of crude oil and petroleum products, changes were under way in the federal agencies. The Atomic Energy Commission (AEC), the Department of Interior (DOI), the Office of Coal Research (OCR), NASA, the National Bureau of Standards (NBS), the Department of Housing and Urban Development (HUD), and NSF began vying for expanded roles in energy R&D. NASA, having cochaired the FCST Solar Energy Panel, aggressively pursued the lead role in the development of renewable technologies and other more esoteric energy technologies such as magnetohydrodynamics (MHD). Under the leadership of NASA Administrator James E. Fletcher, the agency established an energy office, directed by astronaut Harrison H. Schmitt, in order to position itself for leadership in the anticipated major, new technology development programs. This role received sympathetic support on Capitol Hill, where many lawmakers were concerned about NASA's future, now that the Apollo Program had ended. The agency initiated energy projects at the Lewis Research Center, Johnson Space Flight Center, Marshall Space Flight Center, the Jet Propulsion Laboratory, and at the Langley Research Center, where it began construction on a solar-heated addition to one of its laboratory buildings. NASA was confident that, with its engineering skills and high-tech facilities, it could take on this new challenge.

Meanwhile, back at NSF, the RANN program struggled to assert its supremacy in the area of renewable technologies. Without NASA's vast laboratory system to fall back on, RANN chose to showcase its research programs with a series of workshops and conferences to report on progress made by its grantees and contractors. In calendar year 1973, RANN held nine successful workshops and conferences covering all of the solar technology areas listed in table 3.1. More followed in 1974 and 1975, and continued even after most of the RANN energy program responsibilities had been transferred to the newly established Energy Research and Development Administration (ERDA) in 1975. More on ERDA will follow.

The Nixon administration, aware of the budding controversy between NSF and NASA, resolved the issue in June 1973 and named NSF the lead agency to conduct R&D on terrestrial solar energy programs. The importance of this decision cannot be overemphasized. It greatly reduced interagency competition, focused the solar energy research objectives, and at the same time permitted the NASA laboratories to continue their growing involvement in energy R&D, particularly renewable technologies. In early August 1973, James Fletcher sent NSF Director H. Guyford Stever a draft Memorandum of Understanding (MOU) that the staff of the two agencies had labored over for several months. The memorandum acknowledged NSF's leadership in "the development of solar and solar derived energy for terrestrial use," and pledged, "to the extent possible [to] make available to the NSF its personnel facilities and expertise." As an interesting side note, the MOU reserved the right for NASA to conduct R&D in solar and other energy systems for use in space and aeronautical systems. For those who might not remember, it was at this time that NASA began to actively pursue the concept of the solar satellite power system (SPS) as a potentially grand-scale program, larger than Apollo. Although the MOU was never formally signed, the effort entailed in developing it cleared the air, eased growing frictions between the agencies, and promoted close cooperation. The Interagency Panel for the

Table 3.2
Agency membership in IPTASE

National Science Foundation (Chair)
Department of Defense
Navy
Army
Air Force
ARPA
Department of Agriculture
Agricultural Research Service
Department of Interior
National Oceanographic and Atmospheric Administration
Department of Commerce
National Bureau of Standards
Atomic Energy Commission
Agency for International Development
Department of Transportation
Department of Housing and Urban Development
Environmental Protection Agency
General Services Administration
National Aeronautics and Space Administration
Federal Energy Administration
Department of Health Education and Welfare

Terrestrial Applications of Solar Energy (IPTASE) was immediately established, chaired by the author, then a RANN manager, and the first meeting was held in early November 1973. Over the next two years, twenty-one federal departments and subentities or agencies with active interests in solar energy eventually were invited to join IPTASE (see table 3.2); under its umbrella, regular coordination meetings were held over the next six years to oversee the growing number of agency solar projects.

In June 1973, building on its previously submitted budget of $196 million and reflecting an increased awareness of the potential contributions solar energy could make to reduce domestic energy consumption, RANN proposed a greatly expanded solar R&D program to the administration. This new program called for the expenditure of $1 billion over the five years ending in FY 1978. Whereas the earlier $196 million program received immediate support from OMB, the $1 billion program was rejected.

3.4 The Nixon Administration Proposes an Expansion of Energy Programs

Less than one week after the expanded solar program was submitted to OMB, President Nixon announced that he was initiating a $10 billion program for research and development on energy technology to extend over the next five years. On June 29, 1973, the president directed the chairman of the Atomic Energy Commission, Dixie Lee Ray, to "undertake an immediate review of federal and private energy research and development activities ... and to recommend an integrated energy research and development program for the nation."[8] The chairman was directed to report by December 1 her recommendations on the programs that should be included in the president's FY 1975 budget.

After organizational start-up delays, in the fall of 1973 Chairman Ray established sixteen subpanels to review the various, potential energy technology areas and provide the requested recommendations. Subpanel IX, the solar subpanel, was chaired by the NSF's Al Eggers. Because of earlier organizational work that NSF accomplished with IPTASE and the expanded $1 billion June program submission, Subpanel IX was able to rapidly form a consensus among the participating agencies. The subpanel recommended two alternative programs, one designated the "Minimum Viable Program" and the second an "Accelerated Program." The minimum program requested $50 million in FY 1975, for a total of almost $500 million over five years, while the accelerated program, a refinement of NSF's June submission, requested $100 million in FY 1975 and over $1 billion by FY 1979.

At this point, however, a major international event overtook national debate on energy R&D: the OPEC oil embargo, launched in October 1973. Within weeks, President Nixon announced a new energy program called Project Independence, with a stated goal of meeting all the nation's energy needs by the end of the decade (1980) independent of foreign energy sources.[9] This goal was to prove illusory at best and, at worst, ill advised and ill conceived.

Chairman Ray's final report, *The Nation's Energy Future*,[10] submitted to the president in December 1973, bore the clear imprint of politics and agency clout. It endorsed neither of the programs developed by Subpanel IX, instead proposing a much smaller solar program that totaled only $200 million for the five-year period. In response to a request from

AEC for comments on the final draft of the report, Eggers wrote to Richard Pastore, Chairman Ray's staff assistant: "The funding shown ... does not correspond to any real program since the fiscal year totals were apparently arrived at arbitrarily to fit some preconceived funding curve. Therefore, we are not in a position to defend these program funds nor the program objectives which were submitted to us for comment."[11] Despite this protest, the report was submitted with a diminished solar program and with most of the $10 billion initially proposed as a target by the president earmarked for various nuclear technologies. When put into its proper fiscal context, the $200 million solar program represented a smaller program than that approved earlier in 1973 by OMB.

By December 1973, however, support for solar energy R&D was mounting, and release of *The Nation's Energy Future* marked the beginning of the sometimes bitter confrontation between Congress and the administration over priorities for energy RD&D dollars. At one point early in the confrontation, some members of Congress accused the administration of suppressing the Subpanel IX report because the Solar Energy Subpanel recommendations were not included in the formal submission. And when questioned on the low priority given to solar technologies, Chairman Ray was quoted as saying, "Solar energy could be thought of as a flea on the back of the nuclear elephant." This was still an era when some people believed that electric power from nuclear power stations would eventually be so cheap that it could almost be supplied free of charge. The emphasis on nuclear technologies was challenged within the administration by such unlikely sources as Edward Teller, and the projected role of conservation was also disputed as being too restricted. None of these arguments prevailed, however, and the report was submitted with its nuclear bias.

As the OPEC oil embargo persisted through early 1974, the debate on R&D priorities grew in force with a polarization of the competing views of the future energy road the nation should travel. Our dependence on foreign energy supplies for every activity in our daily lives was brought home to all segments of society. In a few months of shortage, the GNP dropped almost $20 billion, 500,000 workers became unemployed and consumer prices increased by almost 10 percent, a third due to higher world oil prices. There was increasing pressure for a quick fix. After all, the argument went, "if we could send men safely to the moon and back, why couldn't we ... ?"

Figure 3.2
Solar Heating and Cooling Transportable Laboratory.

Before discussing the events surrounding Project Independence and its aftermath, however, let us turn to activities at RANN in 1973 and 1974 that provided tangible evidence of the impact solar energy could have. At the end of September 1973, RANN signed three contracts with industry, each valued at over $500,000, the largest RANN contract awards to date. Westinghouse, TRW, and General Electric began multiphased feasibility studies for solar heating and cooling of buildings. Shortly afterward, a fourth major industrial contract was signed with Honeywell to supply a transportable solar heating and cooling laboratory, which RANN planned to take around the country to demonstrate the state of the art of solar heating and cooling technology (see figure 3.2). With a set of roof-mounted solar panels and the latest in Honeywell controls inside the trailer, visitors were to experience space conditioning and hot running water, courtesy of the sun. This mobile display was to prove very popular and made friends for solar energy from Capitol Hill to California. With Westinghouse, TRW, General Electric, and Honeywell under contract, RANN had brought into the program industrial partners who provided instant legitimacy to the NSF agenda.

In addition to these major activities, RANN planned a second large initiative, the Solar Schools Proof of Concept Experiments (POCEs), which would soon provide hard data, some good, some not so good, that shaped the later RD&D approach.

3.5 RANN's Initial, Large-Scale Solar School Projects

In late 1973 RANN senior management decided that a few, large, dramatic solar energy experiments would be valuable, both for the learning experience they would provide and for their public impact. A fast-track Solar Energy School Heating Augmentation Experiment was initiated, and a solicitation invited proposals from existing schools, each teamed with a contractor, to become part of the experiment. Of the many teams that responded, four were selected on the basis of their location, size, type of construction, cost, and an assessment of the interest shown by the local school governing bodies. To accomplish the goals of the four experiments in the time frame announced would require close cooperation, tolerance, and a great deal of patience on the part of local school officials.

Four contracts were signed in record time and in January 1974 design and construction of the experiments commenced. The schools selected were Timonium Elementary School, Timonium, Maryland (north of Baltimore); Grover Cleveland Junior High School, Boston; Northview Junior High School, Osseo, Minnesota (near Minneapolis); and Fauquier High School, Warrenton, Virginia. The experiments involved retrofitting the schools with solar energy systems based on flat plate collectors and hot water storage. Because the solar collector fields ranged from 2,400 ft^2 to 5,000 ft^2, they would represent the largest solar energy systems yet deployed in the United States. System designs required innovative engineering and manufacturing skills because, under the terms of the contracts, the systems had to be installed on existing, operating schools without interrupting school functions and in no more than two months. Three of the four systems became operational by March 1974, and the fourth came on line by April.

The Timonium school system, designed and installed by the AAI Corporation of Baltimore, utilized an innovative, aluminum honeycomb, flat plate collector design (see figure 3.3). The collectors, with a 5,000 ft^2 surface area, were mounted in multiple banks on the roof of the school; a single, 15,000-gallon hot water storage tank was placed at ground level at the rear of the school.

Grover Cleveland Junior High School in Boston was an existing three-story building. Because of the structure's design and lack of space, a large water storage tank could not be placed underground or alongside the building, as at Timonium, or on the roof. Instead, a smaller-than-

Figure 3.3
Photo of Timonium School solar project.

optimum, 2,000-gallon roof tank was included in the design by General Electric, the prime contractor. The roof-mounted collectors were a new GE design, with glass inner glazing, "Lexan" outer glazing to avoid breakage, and integral fluid passages; the collector field was also sized at approximately 5,000 ft^2.

Northview Junior High School's solar energy system, designed and installed by Honeywell, employed flat plate collectors of a "proven" Honeywell design, with an inner "Tedlar" glazing and an outer glass cover (see figure 3.4). The collector field, again about 5,000 ft^2, was deployed on the ground adjacent to the school; the hot water storage tank, restricted by location to a 3,000-gallon capacity, was placed in the school's basement.

The final school system brought on line, at Fauquier High School, was the most innovative of the four experiments (see figure 3.5). The system was designed by InterTechnology Corporation to provide heat to several temporary classrooms, with a 2,400 ft^2 collector field mounted on a separate structure adjacent to the classrooms. The collector were constructed with double-glass glazing and "roll-bond" aluminum plate. At the time,

Figure 3.4
Photo of Northview Junior High School solar project.

Figure 3.5
Photo of Fauquier High School solar project.

the solar collectors were believed to be the largest single planar array in the world. Two 5,500-gallon tanks could store enough heat for up to twelve days of classroom occupancy in overcast weather.

All four systems were heavily instrumented and closely monitored by the contractors. In some, the working fluid, water, contained antifreeze to avoid freezing. Some were self-draining, thus allowing a comparison of the designs for this important operational consideration. Because the four systems did not become operational until late in the 1973–74 heating season, a minimum amount of data was collected, which, however, permitted the contractors to remove most of the "bugs" before the start of the 1974 fall school term.

Over $1.8 million was spent on the School Heating Augmentation Experiment. The performance of the four systems, monitored over several years, varied greatly; the systems were modified from time to time during their operational life, based on the data collected and problems encountered. Because of the need to retrofit in a short time, the economics of the systems were always recognized to be unfavorable. However, much was gained from the experiments, not the least of which was the favorable publicity and the demonstration that large solar energy systems were feasible and had the potential to make sizable contributions if carefully designed. The experiments also demonstrated that U.S. companies had the interest and ingenuity to respond rapidly to a national need when provided with the proper incentives. Perhaps the largest impact was on congressional decision makers who, aware of the operation of the school systems, passed landmark solar energy legislation.

3.6 1974: The Year of Solar Prominence

1974 can be viewed today as the watershed year in the early up-and-down history of the federal solar energy program: 1974 was without question an up year. Activities progressed on all fronts, in the executive branch, Congress, and the research community at large. Optimism was on the rise that, supported by a major solar energy R&D program, solar energy could be a sizable contributor to the nation's energy economy. Earlier predictions that solar energy might contribute 5–10 percent of total U.S. needs by the year 2000 (FCST Solar Energy Panel) were soon to be reinforced by the Project Independence Solar Energy Task Force report.

After announcing Project Independence in November 1973, the Nixon administration was made aware of the enormous task it faced in providing the necessary analysis and in administering the huge resulting program. This and other problems arising from likelihood of continuing energy shortages, with all their related economic impacts, argued for an agency with oversight of all energy matters. Thus the Federal Energy Office (FEO) was established by executive order and began operating in December 1973. At this early date, the administration was also drafting plans for an agency that would consolidate all federal energy R&D, the Energy Research and Development Administration (ERDA), but that would wait in the wings for another year before beginning operations. Six months later, by act of Congress, the FEO was supplanted by the Federal Energy Administration (FEA) and became the focal point for energy policy in the Nixon administration. John C. Sawhill was confirmed as the first administrator.[12]

In addition to developing near-term energy policies for the nation, FEA was charged with collecting and analyzing energy data, evaluating economic impacts of energy programs, and providing regulatory functions for fuels and energy prices. Because the just-completed energy futures study directed by the chairman of the AEC had raised so many hackles, it was decided that the new FEA would coordinate agency inputs to Project Independence and thus provide a more neutral forum for the interagency debates that were sure to occur.

3.7 Project Independence (Blueprint)

Many events, some covered in previous sections, led up to the decision by President Nixon to undertake the major study, Project Independence. Undoubtedly, the Arab oil embargo declared on October 16, 1973, was the final event that made such a study both politically and economically attractive. As we have seen, prior to the embargo, the Nixon administration had undertaken a number of measures to address actual and projected energy problems. With *The Nation's Energy Future* still one month from formal release but already under attack, President Nixon announced Project Independence on November 7, 1973. The goal of this program was stated to be, "... that by the end of this decade we will have developed the potential to meet our own energy needs without depending on foreign energy sources."

Because of uncertainty over where in the federal government responsibility would be placed, the project did not get underway until the spring of 1974. This uncertainty was finally resolved with the formal establishment of the Federal Energy Administration (FEA) on May 7, 1974. With the knowledge that the FEA would soon be in existence, the FEO, called the first government-wide organizational meeting on April 5, 1974. At this first meeting, chaired by Gorman C. Smith, recently transferred to FEA from the AEC, interagency task forces were established for all energy forms. The title of this vast exercise, which eventually drew on the expertise of more than 500 federal employees, contractors, and others in the private sector, had become Project Independence Blue Print, an interesting new distinction. For renewable energy sources two task forces were formed, one for solar energy and one for geothermal energy, the National Science Foundation was asked to lead both. The meeting established a fast-track schedule, calling for a first draft by June and a final report to the president by October 1, 1974. Chairman Smith announced that, with only five and a half years remaining in the decade, a "crash program" needed to be defined to achieve the goals of Project Independence.

The overall objective of the Project Independence study during the next few months evolved to an analysis of four broad policy strategies by which government could reduce or eliminate the need for foreign sources of energy: (1) a "Base Case" strategy, which required only limited new policy actions; (2) an "Accelerated Supply" strategy, which required a large number of key policy actions; (3) a "Conservation" strategy, which was specifically intended to reduce demand for petroleum products; and (4) an "Emergency Preparedness" strategy, which included a major domestic oil storage program. The final report acknowledged that any new government program arising from the study would probably incorporate elements of all four strategies.

Development scenarios for R&D programs, primarily in support of the base case and accelerated supply strategies, were undertaken for each technology area. These became known as "Business as Usual" and "Accelerated" scenarios. For each of these scenarios, gound rules were established, including projected technology market penetrations based on future oil prices of $4, $7, and $11 per barrel. The accelerated program called for major government intervention, including tax incentives, loan programs, special depreciation policies, and other local, state, and federal policies to overcome nontechnology constraints.

The Solar Energy Task Force, led by NSF, drew heavily on the federal personnel, contractors, and other private sector advisors who contributed to the earlier Subpanel IX report as well as the earlier analyses. Although the task force complied with the ground rule prices of $4/$7/$11 per barrel for market penetration analysis, it focused on a projected oil price of $11/barrel, the price eventually reached during the oil embargo, and in the "Overview Assessment" it emphasized the contribution solar technologies could make at this oil price. It was estimated that the various solar energy technologies, developing under the assumptions of the accelerated scenario, could contribute the equivalent of 1.8 quads by 1985 and 7 quads by 1990. These solar contributions were in a projected, total national energy demand of 103 quads in 1985, with continuous growth thereafter. The solar heating and cooling contributions were projected as 0.8 quads and 2.5 quads, respectively. (Out-year estimates for both the "Accelerated" and the "Business as Usual" scenarios can be found in table 4.3.)

These impact estimates, not surprisingly, were very similar to those of earlier studies. Despite the potential impact of solar projected by the Solar Energy Task Force, the summary report of Project Independence chose to downplay solars potential contribution by stating that, beyond 1985, only solar heating and cooling technologies "appear promising" at the higher prices of oil, and that solar power could be a long-term source if economic problems were overcome.

Project Independence strategies concentrated primarily on reducing domestic consumption of oil, thus reducing oil imports, the "hot button" of the day. The R&D programs and technology projections, although the major focus and effort of the various task forces, were not considered a major component of the pre-1985 strategies. While correctly concentrating on the projected energy situation between 1974 and 1985, the Project Independence management acknowledged that, beyond 1985, the rate of introduction of new supply technology would be a key element in holding down oil imports and conserving domestic oil reserves. To some degree, the study avoided some of the pitfalls of *The Nation's Energy Future* by not attempting to consolidate and prioritize the various task force R&D programs. (In fact, R&D budget projections, although the unstated underpinning of these programs, do not appear in the report. The Solar Task Force report, in section 1, "Overview," states that the R&D program initiated follows to a large degree the unpublished, accelerated R&D program submitted by Subpenel IX for *The Nation's Energy Future*, a five-year, $1 billion-plus program, but never mentions a dollar figure.[13]

As a result of Project Independence, the aggressive solar energy program championed by NSF that had been attempting to achieve some official legitimacy for two years was finally available for all to see. The solar heating and cooling component, in particular, was acknowledged as a set of technologies that could make a significant impact as early as 1985, especially if certain incentives could be applied to help stimulate market penetration. For solar advocates, applying incentives to help in the widespread acceptance of solar energy was considered "only fair" in view of the special incentives and government programs that had favored oil and gas production since the turn of the century and that, more recently, had stimulated the development of nuclear energy. Stealing a page from these conventional energy supply sources, the Solar Task Force included an appendix outlining various incentives for both consumers and manufacturers that would be needed to level the playing field for the widespread introduction of solar technologies. Many of these incentives were eventually included in legislation to be passed by federal and state legislatures in the years ahead.

An interesting addition to the Solar Energy Task Force report was Appendix II, "Energy Use and Climate." In 1974 the environmental debate was becoming more heated. Project Independence approached environmental issues somewhat tangentially, recognizing that each strategy would have some impact. For the "Business as Usual" case, the report stated that "environmental impacts are mixed and in some cases lower than 1972"; the "Accelerated Supply" case showed a reduction in air and water pollution and an increase in solid waste. Appendix II was an attempt to closely link use of conventional energy and climate change and to show the environmental advantages of using solar energy. Its analysis went essentially unnoticed.

Project Independence, which early in its history became known as "Project Independence Blueprint" before reverting back to its original title, laid the groundwork for extensive federal intervention into the energy marketplace. Although the final report disclaimed that it was a "blueprint" for reaching zero imports by 1980, the President's original, stated goal, it spelled out the steps needed to approach this goal and to avoid future emergencies through conservation, domestic oil storage, accelerating supply options, and international cooperation.[14] All of these strategies, to some degree, have been accepted in spite of bumps and bruises along the way. Awareness by the public of our total dependence on affordable energy waxes and wanes and parallels to some degree the history of public support for the space program. If there is a problem, long gas lines, if the Russians are beating us to the moon, everyone wants fast action and an immediate solution. If gas is cheap and available, if we win the race to the moon, the public quickly loses interest and the desire to accept changes, especially painful changes, thus laying the seeds for the next emergency.

Notes

1. For further background and references, see K. Butti and J. Perlin, *A Golden Thread* (New York: Van Nostrand Reinhold, 1980), and F. Daniels, *Direct Use of the Sun's Energy* (New Haven: Yale University Press, 1964).

2. U.S. Department of the Interior, *Manual on Solar Distillation of Water: Progress*, Report no. 546, 1970.

3. The President's Materials Policy Commission, *Resources for Freedom*, June 1952.

4. Richard M. Nixon, "A Program to Insure an Adequate Supply of Clean Energy in the Future," June 4, 1971, as reprinted in *Executive Energy Documents* (Senate Committee on Energy and Natural Resources, July 1978), 1–12.

5. NSF/NASA Solar Energy Panel, *Solar Energy as a Natural Resource*, December 1972.

6. U.S. Congress, House of Representatives, Committee on Science and Astronautics, *Report of the Task Force on Energy*, December 1972.

7. National Academy of Engineering, Committee on Public Engineering Policy, *Priorities for Research Applicable to National Needs: Report of the Ad Hoc Steering Committee for the Study of Research Applied to National Needs*, 1973.

8. Richard M. Nixon, "Statement on Energy," June 29, 1973, as reprinted in *Executive Energy Documents* (Senate Committee on Energy and Natural Resources, July 1978).

9. Richard M. Nixon, "Address on the Energy Emergency," November 7, 1973, and "Address on the National Energy Policy," November 25, 1973, as reprinted in *Executive Energy Documents* (Senate Committee on Energy and Natural Resources, July 1978).

10. *The Nation's Energy Future*, Report to Richard M. Nixon, President of the United States, December 1973, Wash-1281.

11. A. J. Eggers, Jr., Assistant Director for Research Applications, NSF, memorandum to Richard Pastore, Staff Assistant, Energy Reorganization Unit, AEC. Subject: NSF Comments on the Draft of Chairman Ray's December 1, 1973 "Report to the President on Energy Research and Development," 1973.

12. For additional detail see R. M. Anders, *The Federal Energy Administration* (U.S. Department of Energy, 1980).

13. Federal Energy Administration, *Project Independence Blueprint Final Task Force Report: Solar Energy*, November 1974.

14. Federal Energy Administration, *Project Independence: A Summary*, November 1974.

4 The Buildup Years: 1974–1977

Donald A. Beattie

4.1 Congressional Activities

Well before the oil embargo in October 1973, and spurred in part by the Nixon administration, the 93rd Congress began to debate the shape of the nation's energy future. Concern about the future was bipartisan even though the leadership in both houses of Congress was controlled by the Democratic Party. As a result of this concern, seven bills that changed the renewable energy landscape were passed by the 93rd Congress and signed into law. These were Public Law (PL) 93-409, the Solar Heating and Cooling Demonstration Act; PL 93-410, the Geothermal Energy Research, Development and Demonstration Act of 1974; PL 93-383, the Housing and Community Development Act of 1974; PL 93-438, the Energy Reorganization Act of 1974; PL 93-473, the Solar Energy Research, Development and Demonstration Act of 1974; PL 93-577, the Federal Nonnuclear Energy Research and Development Act of 1974; and PL 93-275, the Federal Energy Administration Act. (For additional background and detail on these laws, see chapter 1 of volume 10 in this series.)

The seven laws covered the full spectrum of energy concerns. In particular, Congress, along with the administration, acknowledged the lack of a focal point in the executive branch for energy matters; this was addressed in PLs 93-275 and 93-438. The concern about future energy supplies led to PLs 93-383, 93-409, 93-410, 93-473, and 93-577, with special attention given to nonconventional energy sources; renewables, and especially solar energy, received the bulk of the attention. Because the concern was bipartisan, and because the laws largely reflected administration energy philosophy, they were rapidly implemented. In the following sections, we will discuss these laws, how they affected the ongoing federally supported solar energy programs, and the agencies involved. To help the reader place in context the changing agency participation and management responsibilities discussed in this volume, figure 4.1 provides a simplified guide to these roles, plotted against time.

As interest in solar programs grew, and as senators and congressmen and their staffs saw an opportunity to show the folks at home they were actively involved, NSF was snowed under with correspondence from the

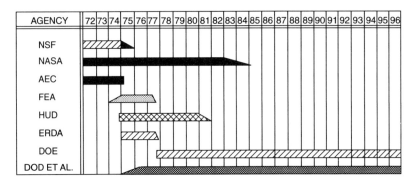

Figure 4.1
Agency participation time line.

Hill. Every member of Congress pressed to have a demonstration sited in his or her district or state. In the halls of NSF the "magic number" for federally funded demonstrations became 535 (one for each senator and congressman), despite the numbers called for in the various bills being debated. Added to this correspondence was that of legitimate, and not so legitimate, proposers, inventors, and just plain citizens wanting to make a contribution. The small NSF staff, numbering about 17 professionals by 1974, was overwhelmed. Answering the correspondence left no time to do the "real work" of the office. Telephone lines were ringing continuously. To maintain his sense of humor, one manager kept two separate piles of phone messages to return: one pile for calls that came in while he was away from his desk, another was for calls that came in while he was on the phone. Both piles were usually just as high.

Part of the growing correspondence load was caused by the need to supply background or "draft" letters for those in Congress and the White House who were also trying to respond to constituent mail. Not responding to a congressman, senator, or White House staffer in a timely fashion was a real no-no. White House requests usually had a 3–5 day turnaround after being logged in at NSF. All told, the correspondence load from Congress and the White House would often exceed 100 requests per week. A long concurrence chain preceded the release of NSF replies to those requests, and there was the constant fear that replies would get lost somewhere in that chain.

Because most correspondence could not be adequately answered with form letters, the rapidly rising level of paperwork of all types led to a struggle to expand staff. A person was hired just to answer phones, with no other secretarial duties. Faced with a ceiling on Federal employees; detailees were brought in, as well as people on sabbatical from universities who had helped on peer reviews and had knowledge of the programs. It of course did not help that NSF, housed at the time at G Street Northwest, two blocks from the White House, had restricted floor space to accommodate growth. Little by little, all these difficulties were dealt with (although never completely overcome), and the volume of angry reprimands from the White House and Congress abated.

Responding to correspondence was obviously just one of the many problems involved in managing a rapidly expanding very visible, R&D program. Budgets were more than doubling every year. Program and budget planning cycles were becoming increasingly intricate, as new congressional committees assured more and more oversight over the program. Program content was also becoming more complex—a natural evolution as additional funding made it possible to undertake supporting research in many new areas.

These conditions led to an interesting confrontation in early 1974 between NSF's Al Eggers and Congressman Mike McCormack (D-WA), who chaired an energy task force for the House Committee on Science and Technology. McCormack was viewed as a friend of solar energy and a strong advocate of bringing new energy sources into the supply mix. The NSF Solar Schools POCEs, just under way, had caught his eye, and he undoubtedly thought that if four POCEs were good, thousands would be better. But when asked to testify on proposed legislation that would greatly expand solar demonstrations, Eggers suggested to McCormack that it was too soon to take on such a vast program, citing as two of his reasons the lack of maturity in the technology and the shortage of management resources. These hearings, beginning in late 1973 and carrying over into 1974, heralded the vigorous push by Congress to pass solar legislation.

4.2 International Energy Cooperation

The concern for future energy supplies was worldwide, and the rapid expansion of the U.S. solar energy program led a number of countries to rethink their own energy R&D programs. Solar energy in all its forms was attractive, especially for countries with a heavy dependence on imported oil. This interest led to a number of bilateral and multilateral science and technology exchanges that concentrated heavily on solar energy technologies.

4.2.1 NATO Commission on the Challenges of Modern Society (CCMS)

In 1969 President Nixon proposed establishing the CCMS in order to provide a "social dimension" to NATO. The commission would serve as a forum in which member states would share their economic expertise and technical know-how to respond to global challenges such as air and water pollution, food and raw materials shortages, and the need to maintain energy availability. A lead country was designated for each issue to be analyzed, with other countries participating at a level of their own choosing.

A solar energy pilot study was initiated in November 1973. The United States was designated the lead or "pilot" country, with France and Denmark as "copilots." Twenty-two additional countries (including some outside of NATO) and UNESCO eventually participated in this pilot study, which undertook a broad-ranging investigation of cost-effective and practical applications of solar energy to heating and cooling in residential, commercial, industrial, agricultural, and public buildings. The formal title of the pilot study was "Solar Heating and Cooling Systems in Buildings." Raymond H. Fields of NSF was designated project director, and Fred Morse chaired the study.

A primary objective was to encourage information exchange among the participating countries through reports and meetings on the progress being made on national solar projects. Each country selected the projects it wished to pursue, although some specific joint projects, such as a "zero energy house," were also undertaken. Three countries had projects underway in 1974, and one year later thirteen countries had projects on which to report at the annual meeting.

The U.S. program was by far the largest and incorporated all of the solar heating and cooling demonstrations supported by the NSF, and later ERDA and DOE, as well as demonstrations overseen by other government agencies. The total list of demonstrations approached 100 distinct projects by 1975. In addition, results of research projects ranging from thermal storage to collector design to control system modeling became a part of the information exchange shared by the U.S. team.

The CCMS solar energy pilot study helped pave the way for the more broadly based International Energy Agency (IEA) and demonstrated that international cooperation was possible in addressing energy problems, even when different countries looked at such problems and their solutions in quite different ways. The cooperative spirit of the CCMS Solar Pilot Program carried over to the IEA Solar Heating and Cooling Programme, which continues actively today with twenty participating countries.

4.2.2 U.S.-U.S.S.R. Cooperation in Solar Energy

In May 1972, at about the same time that cooperative solar energy projects were beginning with the NATO countries, President Nixon signed the "Agreement between the Government of the United States of America and the Government of the Union of Soviet Socialist Republics on Cooperation in the Field of Science and Technology," which led to the establishment in 1973 of the U.S.-U.S.S.R. Joint Commission on Scientific and Technical Cooperation. This commission in turn agreed to focus on nine energy areas, one of which was "General Technology for the Utilization of Solar Energy." H. Guyford Stever, who was then also the President's Science Advisor, signed for the United States. Because at that time NSF was lobbying the administration to assume the lead role among government agencies for solar energy R&D, Stever assigned RANN Deputy Assistant Director Richard J. Green, to be the U.S. coordinator for solar projects. Negotiations began immediately to arrange for an exchange of "specialists" before the end of 1973.

A number of events, including the oil embargo, prevented an exchange in 1973, and thus it was only in January 1974 that Green and a small delegation traveled to Moscow to sign the protocol that would initiate the first exchange of "specialists" in solar energy. By this time, NSF's solar programs were beginning their explosive expansion. Soviet activities in solar energy were well known and reported in national and international publications; it was believed that Soviet scientists had made important strides since the 1950s. Some U.S. researchers who had visited the U.S.S.R. in the intervening years had viewed and discussed projects with their Soviet counterparts. Thus the basis of the agreement and subsequent protocols was that activities undertaken would be of "mutual benefit, quality and reciprocity." It was expected that the U.S. side would gain as much from the joint projects (to be defined) as the Soviets.

Although this period represented a thaw in relations with the Soviet Union, it was still difficult to arrange exchange visits. Itineraries and schedules changed constantly, and it was especially difficult to get agreement that the content of the itineraries would be of equal "quality." The first Soviet team of four "solar specialists," led by Dr. Ivan T. Alad'yev, U.S.S.R. Coordinator for Solar Energy Projects from the G. M. Kazhizhanovsky Power Engineering Institute in Moscow, arrived in the United States in mid-June 1974 and visited a variety of government and university laboratories as well as some government-sponsored industrial projects during their two-week stay. They seemed most impressed, however, not by the advances we were making in solar technologies but by a trip to Disneyland that was squeezed in while the visitors were on the West Coast.

The reciprocal visit to the Soviet Union by a U.S. delegation of eleven researchers and government managers took place in September 1974. The morning after its arrival in Moscow, the delegation was informed that a prime stop on the itinerary, the Physical-Technical Institute in Ashkhabad, directed by Dr. V. A. Baum, one of the best-known Soviet researchers, had been canceled. The reason given was that the airport was closed and would not reopen before the visit ended. Amid considerable grumbling from the U.S. delegates, the schedule was juggled, and the team went on to visit institutes in Moscow, Tashkent, Bukhara, Yerevan, and Baku, with a final joint seminar in Moscow.

During the visit, it became clear that there were few new or developing projects at any of the institutes. Most laboratory equipment was outmoded, and it was evident that funding of solar research did not have a high priority. At many sites the equipment was not fully operational and, in a few instances, seemed to have been dusted off for the visit after a long period of neglect. Despite Moscow's claim that the Soviet solar program was subject to central planning and oversight, on-site discussions confirmed that each institute was on its own and had to scramble for support. As noted in the official U.S. trip report: "Given the poorly defined future Soviet efforts, it is difficult to obtain matches between national efforts that would be extremely useful to both sides."

In spite of this negative assessment, cooperation continued sporadically for six years, until the beginning of the Reagan administration, when relations between the two countries became less than cordial. Although attempts were made during those six years to conduct joint seminars and to undertake projects that included research on the development and testing of "solar air-conditioning and heating systems," the projects never came to fruition, and the seminars were not actively pursued.

4.2.3 Other International Programs

This period saw a number of bilateral agreements between the United States and other countries. The cooperative science and technology agreement with France involved the National Center for Scientific Research (CNRS). At that time, France had by far the world's largest operating solar furnace (1,000 kW) at the Solar Energy Laboratory in the Pyrenees Mountains. The agreement made it possible for NSF grantees and contractors to test and evaluate components and subsystems that would eventually find their way into U.S. programs such as the "Power Tower" project at Sandia Laboratory in Albuquerque, New Mexico.

The bilateral agreement with Saudi Arabia led to a long-term cooperative relationship under the Joint Commission on Economic Cooperation. Whereas most of the bilateral energy agreements were predicated on the idea that the United States would gain as much technologically as the foreign partner, it was recognized that the information flow with the Saudis would be mostly one-way, with the reward for the United States being the accumulation of design, construction, and operating experience that would result from the jointly funded (50/50) projects. The agreement was also looked upon as a vehicle by which U.S. companies could enter the Saudi marketplace, which was projected to be extremely attractive in view of the booming Saudi economy resulting from sharply higher oil prices. The Saudis, meanwhile, realized that their domestic energy needs were growing rapidly and their vast petroleum reserves would one day be depleted. Maintaining as much oil in the ground as possible by reducing their own consumption would eventually be to their benefit. The Saudis saw their geographical location as fortuitous, allowing them to operate solar energy systems very economically due to the high daily solar flux; U.S. know-how and technology expertise would help them become knowledgeable in this expanding field. The resulting joint program, called

"SOLERAS," provided $50 million from the Saudis for a wide variety of demonstration projects.

Other bilateral agreements with Australia, Israel, Italy, and Japan contained specific energy provisions. The Japanese, after visiting NSF in 1974, announced in 1975 an aggressive, wide-ranging solar energy research program called "Project Sunshine," which was projected to spend the equivalent of several billion dollars over the next twenty-five years. A bilateral agreement for the exchange of science delegations made between the National Academy of Science and the Science and Technology Association of the People's Republic of China led to a continuing low-level, energy technology exchange that ebbed and flowed with the changing political climate.

Many U.S. companies, especially those with foreign subsidiaries in the above countries, participated in the resulting programs, including Honeywell, General Electric, Heliotek, and Westinghouse. In general, because of the rapidly increasing investment by the U.S. government in solar energy R&D, U.S. participants in these agreements felt that American technology was becoming preeminent. The one area of exception was solar-powered water heaters, where Australia, Israel, and Japan had existing industries that were supplying and servicing thousands of units yearly.

4.3 The Ford Foundation's Energy Policy Project

For the twenty-three years leading up to the oil embargo of 1973, U.S. energy consumption had grown at an average annual rate of 3.4 percent. And in the eight years immediately preceding the embargo, energy consumption grew at an average annual rate of 4.5 percent. Most energy "experts" believed that this historical growth rate in energy consumption was the underlying engine that drove the U.S. economy and controlled the growth of the gross national product (GNP). Common wisdom taught that GNP could not grow without an increase in per capita energy usage. Most attempts to project energy futures simply assumed a thirst for ever larger amounts of energy and tried to devise new supply alternatives, either through conventional means or by bringing new sources on line. At NSF a few RANN studies stood out for supporting the idea that adoptation of energy efficiency measures, especially in the nation's building sector, could markedly reduce the growth of energy usage in the years ahead.

In December 1971 the Ford Foundation authorized the Energy Policy Project. S. David Freeman, former special consultant on energy to the Senate Commerce, Science, and Transportation Committee (and future Energy Advisor to President Carter), chaired the project. At its start, the project attracted little notice; although the Nixon administration was actively pursuing new energy initiatives in response to annoying energy shortages, there was no crisis atmosphere to stir attention. By the time the report was published in October 1974, the situation had changed significantly.

The Energy Policy Project[1] closely examined the relationship between jobs, the GNP, and energy growth. It established three scenarios: (1) historical growth, (2) technical fix, and (3) zero energy growth as a means to bound the tie among energy, jobs, and GNP. The historical growth scenario examined the effect of continuing energy consumption growth of 3.4 percent annually through the year 2000. The technical fix scenario projected a lessening in the rate of energy consumption growth to 1.9 percent from 1974 to 2000, although such a low energy growth rate had never occurred over any sustained period in the country's history. The technical fix accomplished this low growth rate through the application of energy conservation technologies (1) at the point of energy consumption, such as improved automobile fuel economy and increased building insulation; and (2) at the point of energy production, such as improved operating efficiencies at power plants and refineries. And, as might be expected, zero energy growth (ZEG)—the most radical scenario—called for adoption of a great many changes, both technological and sociological. As an example of the project's "radical" thinking, most of the savings in the transportation sector would come from mandating average fuel economies for the private automobile of 20 MPG in 1985 and 25 MPG by 2000. Such economies were projected to save 5.9 quads by 1985 and 9.9 quads by the year 2000.

Under the zero energy growth scenario, energy use would continue growing, albeit at a rate lower than the historical rate, until 1985, when it would reach approximately 100 quads (from an actual 75 quads in 1973); because of a large number of life-style and end-use improvements stimulated by government incentives or disincentives, it would then level off. Somewhat surprisingly, solar energy was not expected to be a major factor in this scenario, which took the same track as the technical fix scenario, even though the Ford project, like all the studies of this era, acknowledged that solar heating was the most advanced solar technology. The project predicted that solar energy might contribute up to 1 quad by 1985 and 2–4 quads by the year 2000 and that the key element in achieving zero growth would be energy conservation.

The Energy Policy Project report was attacked from opposing sides. For energy supply siders, it was heresy to think that the economy could be sustained at zero energy growth, or even at a growth rate of 1.9 percent per year. For renewables advocates, any energy future that did not show major substitution of solar energy for conventional energy by the year 2000 was beyond belief. For energy conservationists and the growing environmental lobby, however, the report showed the way to the future. And the fact is, the Energy Policy Project proved more than prescient. Market forces and government actions have led individual and industrial consumers to make a number of wise choices, and the U.S. energy economy has quickly adapted to growth without increased energy consumption. Even today, while energy prices are lower in terms of 1973 dollars than they were at the beginning of the energy crisis, there appears to be little pressure or need to return to precrisis growth rates. We have been living with zero energy growth for almost twenty years. From a solar energy perspective, however, the effect of low growth in energy demand and low, relatively stable energy prices has been to limit, severely, market niches for the introduction of solar technologies.

4.4 A National Solar Energy Program Is Submitted to Congress

At the end of 1974, after a busy year both on Capitol Hill and at the White House, NSF submitted a five-year national solar energy program to the House and Senate, as required by the House conference report on PL 93-413, the National Science Foundation Authorization Act of 1974. This program attempted to capture the key elements of the FCST Solar Energy Panel report, the Subpanel IX report, Project Independence, and the five-year plan submitted by NSF to OMB in June. The funding levels and program content were consistent with the June submission but acknowledged the changing solar RD&D situation brought about by the flurry of legislation enacted by the 93rd Congress.

Of particular importance was PL 93-409, the Solar Heating and Cooling Demonstration Act, which required HUD and NASA to submit a plan to accelerate the demonstration of solar heating and cooling systems on residential buildings. The NSF five-year program was held to be up to date in all areas except heating and cooling of buildings, where modifications would be made based on the HUD/NASA plan. HUD, under the leadership of Joseph Sherman, and NASA, led by Donald Bowden of the Marshall Space Flight Center (MSFC), Huntsville, Alabama, developed a plan that would require $60 million spread over five years. Both HUD and NASA would be responsible for procuring and demonstrating residential and commercial systems, the latter for multifamily residences. The early demonstrations, selected through Requests for Proposals (RFPs) to be released in the spring of 1975, would emphasize available, off-the-shelf systems. It was anticipated that 300–500 units would eventually be installed on the basis of these RFPs. In addition to the demonstrations, NASA planned on spending almost $20 million of the $60 million developing new systems. HUD agreed to monitor system performance and disseminate data and information on the demonstrations.

The HUD/NASA plan was submitted to Congress on December 30, 1974, a fitting ending for the year of solar prominence. With the advent of ERDA and with all the new solar legislation signed into law, 1975 would be a busy transition year as the executive branch agencies struggled to consolidate and expand the accomplishments of 1974.

4.5 Establishment and Impact of ERDA

On October 11, 1974, after the 93rd Congress had closed its books and gone home to campaign, President Ford signed PL 93-438, the Energy Reorganization Act of 1974, initiating a new era in government-sponsored energy RD&D. This wide-ranging act abolished the Atomic Energy Commission and created three new organizations: the Energy Research and Development Administration (ERDA), the Nuclear Regulatory Commission (NRC), and an Energy Resources Council made up of senior cabinet-level and agency managers. The establishment of ERDA had the most far-reaching effect because it brought together under one administrator energy RD&D programs previously assigned to several departments and agencies, including the AEC, NSF, Department of

Interior, and EPA. President Ford chose Robert C. Seamans, Jr., then president of the National Academy of Engineering, to lead ERDA.

Seamans took office in December, prior to the formal establishment of ERDA, and immediately set up shop at NSF. He began interviewing candidates to fill the senior management positions called for by the act, including six presidential appointees to fill assistant administrator positions. John M. Teem, AEC Assistant General Manager for Physical Research, was selected as Assistant Administrator for Solar, Geothermal, and Advanced Energy Systems. His office was designated to encompass the NSF and AEC solar and geothermal programs as well as the former AEC's Controlled Thermonuclear Research (Fusion) and Physical Research programs. Upon the formal establishment of ERDA on January 19, 1975, Teem, while awaiting confirmation, also established an office at NSF.

The new ERDA was to bring together over 7,000 federal employees from agencies that, especially in the solar and geothermal programs, had just recently been friendly, or not so friendly, rivals for federal research funds. Teem quickly established solar and geothermal task forces to meld the NSF and AEC programs and start this difficult transition. Although the total dollar value of the programs was less than $70 million, the interactions were complex. NSF and AEC had operated under very different cultures. Because NSF's traditional role was to support basic research, its management and project selection criteria differed markedly from AEC's, and the majority of the solar and geothermal funds and programs transferred to ERDA came from NSF (totaling some $51.7 million). ERDA's new management procedures, modeled primarily after the AEC's, were much more constrained.

Beginning in June 1974 and at the request of OMB Associate Director Frank G. Zarb, NSF reviewed its energy programs to determine which should be transferred to ERDA and which would stay. Under the guidelines issued by OMB, after ERDA reached full operational status, NSF would still continue to support cutting edge energy R&D, but at a reduced funding level. The areas fenced off for NSF, under RANN management, included longer-range, higher-risk, and interdisciplinary research. NSF would be excluded from doing full-scale working demonstrations, such as the school POCEs. NSF Director H. Guyford Stever clarified this distinction in correspondence with Zarb and stated that, based on hearings before the Congress, "It has clearly been the intent of the Administration and I believe Congress for NSF to continue its research support role on a range of energy activities."

This exchange resulted in a tug of war of definitions—what was basic research, what was applied, what was a demonstration? Not sure where their future lay, NSF staff defined these terms broadly, which enabled NSF to justify continuing a fairly large program that included almost half of the ongoing projects. Solar programs totaling $20 million were identified for transfer, with $10 million allotted to the Solar Heating and Cooling of Buildings program alone. The $20 million total budget met the OMB target requested by Zarb. Eventually, as roles and missions continued to be debated, NSF transferred $37 million of FY 1975 solar energy program authority to ERDA, almost double the OMB target figure, with the understanding that additional programs were planned for transfer. In addition, almost the entire NSF professional and secretarial staff managing the solar program, some twenty people, were transferred to ERDA.

Congressional interest remained high. There was continuing concern that solar programs would be swallowed up in the new ERDA bureaucracy and, more darkly, a bureaucracy dominated by former AEC staff. On January 28, 1975, ten days after ERDA's formal start-up, Thomas P. Ratchford, staff director for the Energy Subcommittee of the House Science and Astronautics Committee, requested a meeting with the new ERDA staff to make known the concerns of the committee, in particular, to determine how the NSF/ERDA transfer would be organized. Ratchford also reminded the ERDA staff that, in keeping with the recently passed legislation, the committee wanted to continue an aggressive demonstration of 4,000 housing units incorporating solar energy systems. In April 1975, in response to questions by Chairman Olin E. Teague of the House Committee on Science and Technology, NSF Director Stever wrote that it was "Administration policy ... to maintain the momentum of the solar energy program," and that he understood from Seamans that, based on the earlier planning of NSF, "the overall national solar energy plan is nearing completion under ERDA leadership."[2] He ended his letter by saying, "I believe we have come a long way in a short four years at the beginning of which terrestrial solar energy research was a minor item in the NSF/RANN budget to where it is now a major thrust of ERDA at two orders of magnitude increase in the budget." Finally, he offered to meet with Chairman Teague along with Seamans, Teem, and Eggers, "to clear up any remaining issues of concern regarding the solar energy program." In spite of these words of assurance, Congress remained skeptical and suspicious in the months and years ahead.

By March 1975, 172 solar and geothermal proposals, the majority dealing with solar programs, had been transferred from NSF to ERDA with the understanding that most would be funded. However, the solar proposals, primarily in the area of solar heating and cooling, were about to be put in competition with greatly expanded, congressionally mandated programs. Under the National Science Foundation Act of 1950, PL 81-507 (64 stat. 149), NSF had broad and unique authority through which it could fund proposals. The Foundation could provide grants that, by their unique nature, required minimal oversight and reporting; it utilized program solicitations, a relatively simple mechanism, to provide a competitive environment, and it also entertained and funded "unsolicited" proposals. This broad mix of proposal types, some received from established NSF researchers, was now subject to a new, more tightly constrained ERDA approach. In recognition of these problems, Assistant Adminstrator Teem asked for, and received, a "deviation" from the ERDA procurement regulations to allow him to more closely observe NSF award procedures during the transition.

The final chapter in the story of the transfer of energy programs from NSF to ERDA was written at the end of 1975. In six months, OMB had radically changed its position on NSF's role in energy RD&D. Based on the guidance and budget mark supplied to NSF for FY 1977, RANN came to the conclusion that, with the exception of energy systems analysis, all energy RD&D programs would have to be transferred to ERDA. This would include approximately $5 million of funded programs in solar and geothermal energy. OMB's sudden turnabout had some potential for embarrassment because NSF had continued to accept proposals in many energy areas and had on hand at the end of the year some $15–20 million of proposals that now had no chance of receiving NSF funding. The ERDA and NSF staffs jointly reviewed this backlog and identified proposals that were of interest and that could be accommodated in the ERDA budget. For the remainder, NSF had the unpleasant task of notifying proposers that NSF energy programs had come to an end.

Despite the sad demise of the NSF solar energy programs, their influence was beyond question. NSF had not only laid the groundwork for the planning and execution of a rapidly expanding solar RD&D program, but had established a set of simplified procurement procedures that permitted small business and individual entrepreneurs to participate in the solar program. As noted above, Assistant Administrator Teem early on received a deviation in procurement regulations to allow ERDA to continue along the lines of NSF procurements. This experience proved very successful and, because most of the ERDA solar staff had come from NSF, ERDA management was encouraged to be innovative. This led to the development and use of Program Opportunity Announcements (POAs) and Program Opportunity Notices (PONs), which significantly reduced the paperwork required of proposers to get their ideas before ERDA management.

Another early area of contention was ERDA patent policy. This was based on AEC policy, which generally reserved for the government patents resulting from government-funded work. In the nuclear energy world, this was seldom problematic, but for ERDA it became a major problem, especially in solar RD&D, where the hope of commercialization was attracting small businesses to the field. NSF's patent policies were much more flexible, allowing proposers to negotiate patent and intellectual rights positions with the government. ERDA General Counsel R. Tenney Johnson, who came from a more liberal NASA patent background, initiated a review of ERDA patent policy in the summer of 1975. John Teem became a leading advocate for a progressive patent policy and a special champion of allowing ERDA's RD&D industry partners to protect their background rights. A new, more liberal patent policy was finally promulgated but encountered considerable resistance. Most of ERDA's contract negotiators, both in the field and at headquarters, came from the AEC and had a hard time giving up the principle that the government should have the background rights and patents for government programs undertaken with industrial partners. It was many years before this culture was eliminated and industry felt comfortable negotiating with ERDA and its successor, DOE.

4.5.1 ERDA's First National Plan

PL 93-577, the Federal Nonnuclear Energy Research and Development Act of 1974 required ERDA to develop a comprehensive plan for energy research, development, and demonstration. The plan was to be designed so that the country could achieve solutions to energy supply and associated environmental problems in the short term (early 1980s), middle term (1980s to 2000), and long term (beyond 2000). The first plan was to be submitted by the end of June 1975, and updated every year thereafter. The need to respond to this requirement in less than six months placed a great burden on Administrator Seamans and his skeletal staff because he was still selecting his senior managers and getting the disparate groups he had inherited to work together as a team. To add to his difficulties, he was required to consult with many other federal departments and agencies, as well as the private sector, during the preparation of the plan. Seamans gave Roger W. A. Le Gassie, his new Assistant Administrator for Planning and Analysis, the job of putting the plan together.

Working (often quite literally) around the clock, Le Gassie, supported by his fellow assistant administrators as they came through the slow process of formal Senate approval, met the deadline, and the plan, designated *ERDA-48*, was submitted to the House and Senate on June 28, 1975.[3] To make the deadline, only volume 1, *The Plan*, was submitted at that time, and volume 2, *Program Implementation*, a much larger document that contained the detailed description of each of the nuclear and nonnuclear programs, was submitted some months later. *ERDA-48* was an interesting mix of old and new, the old being the programs advocated by *The Nation's Energy Future* and Project Independence, and the new characterized by a growing awareness of the role of conservation and the potential of "underused technologies" and inexhaustible energy sources to make significant contributions in the near to long term.

Once again, as in the earlier studies, multiple scenarios with differing assumptions were analyzed to determine the sensitivity of energy supply/demand forecasts to various remediation strategies. In his forwarding letter, Seamans stated that "the conclusions contained in the plan confirm the urgent nature of the energy challenge confronting the nation." This challenge was defined as the heavy reliance on imported energy, which accounted for 20 percent of the total domestic energy consumption, and the related national security implications. Added to these concerns was the quadrupling of the price of oil, the primary source of imported energy, in just two years' time, which, if continued, was projected "to threaten U.S. economic stability."

ERDA-48 analyzed six scenarios, named after their primary characteristic: (0) "No New Initiatives"; (I) "Improved Efficiencies in End Use"; (II) "Synthetics from Coal and Shale"; (III) "Intensive Electrification"; (IV) "Limited Nuclear Power"; and (V) "Combination of All Technologies." Each scenario was examined for its short-, mid-, and long-term impact on reducing the need for imported oil and gas. For each scenario an estimate was made of the contribution that could be expected from new technologies, including conservation techniques, in meeting projected national energy demands. In early 1975, the plan estimated energy demand in the years 1985 and 2000 to be 95–105 quads and 115–165 quads, respectively. Again, remember that actual energy consumption in 1974 was 73 quads; the inevitability of growth in energy consumption was taken as a matter of faith. From the perspective of solar energy, scenario IV, "Limited Nuclear Power," was estimated to provide the greatest opportunity for solar to make an impact: 5.25 quads by 1985, and a whopping 103.5 quads in the year 2000. Scenario III also projected major contributions from solar energy, while scenario 0, "No New Initiatives," gave solar energy zero impact in both 1985 and 2000.

The authors of *ERDA-48*, based on the above analysis, went on to rank and prioritize the RD&D programs for each technology. Solar electric technologies were ranked with the highest-priority supply technologies, while solar heating and cooling was included in a group called "Other Important Technologies." On the surface, this ranking was encouraging to ERDA's small Solar Division staff, but the reality of being ranked high was that the solar RD&D programs were put into direct funding competition with high-profile, long-established programs such as the breeder reactor, fusion, and synthetic fuels. However, *ERDA-48* assured that solar energy would at least get invited to the table when the budget pie was divided. Considering that former AEC staff now controlled ERDA's budget process, this ranking would prove to be an important first step. Two years later, as discussed in section 4.9, the picture would change, and solar energy RD&D would be accorded a low priority by ERDA management.

While *ERDA-48* was being prepared, the International Solar Industry Expo 75, sponsored by the Solar Energy Industries Association (SEIA), ERDA, and FEA was held in May in Washington, D.C. In a preamble to the three-day program, President Ford wrote, "The goal of this nation's solar energy program is to develop and introduce at the earliest feasible time those applications of solar energy that can be made economically attractive and environmentally acceptable as alternative energy sources." He ended his preamble by stating, "I pledge my continuing support to these efforts." Such proactive language came as a pleasant surprise to the solar community because it represented the first time that any administration had gone on record with a strong endorsement for solar energy.

The Solar Expo was very successful, attracting thousands of attendees from industry, government, and academia. On the final day, SEIA presented Congressman Mike McCormack with its first "Man-of-the-Year" award.

4.5.2 The National Solar Energy Research Development and Demonstration Program

While ERDA was developing its first national energy plan (*ERDA-48*), the Solar Division was leading a government-wide effort to construct a new national solar energy RD&D program (designated *ERDA-49*). Such a plan was mandated by PL 93-473, the Solar Energy Research, Development and Demonstration Act of 1974. NSF's national solar plan, as submitted to Congress in December 1974, was based primarily on inputs to two major studies but was constructed in the absence of any overall economic or environmental restrictions. Or, to put it in a different context, NSF's solar plan assumed, optimistically, that there would be essentially no constraints on the entry of solar technologies into the energy marketplace, regardless of potential competing energy forms.

With the PL 93-473 requirement as a club, Chairman McCormack called a special hearing of his subcommittee in May 1975 to determine the status of "The Plan." ERDA'S John Teem was the principal witness and received the brunt of the criticism from the chairman and other subcommittee members, who felt that ERDA was dragging its feet, even though they testified that the national solar energy plan was nearing completion. It was a bipartisan attack, with Congressman Barry M. Goldwater Jr. (R-CA) also taking Teem to task for ERDA's slowness in providing a plan. A clear message was sent—release the plan ASAP.

In response to the congressional urging, *ERDA-49* was issued in June 1975.[4] The plan attempted to take a more realistic approach than NSF had by staying within an overall framework established for the analysis of all energy technologies in *ERDA-48*, which had selected a set of eight specific goals to shape the framework of a national plan for energy RD&D. The solar planners believed that solar energy technology, aided by a large federal investment in RD&D and private sector investments, could make a major contribution to half the *ERDA-48* goals. As such, solar had to be considered a major actor in the future mix of energy supply technologies. Knowing congressional support to be strong, and projecting ever larger RD&D budgets into the future based on President Ford's Solar Expo 75 statement, ERDA planners quickly succumbed to the optimism found in earlier plans.

ERDA-49 mapped out a broad strategy that included the rapid deployment of technology demonstrations supported by the government's RD&D investments, which would lead to pilot-scale facilities. By taking this approach and closely involving the private sector, it was felt that market barriers could be quickly assessed and overcome and, where necessary, federal incentives could be developed to bypass institutional road blocks brought about by unfamiliarity with the new technologies. Although unspecified in the plan, the foundation of this ambitious program was an RD&D budget greater than $1 billion spread over the next five years.

ERDA-49 attempted to address the complete range of problems that could prevent solar energy from becoming a major energy contributor. Because each solar technology was acknowledged to be at a different stage of development, different efforts and directions would have to be followed for each technology. The emphasis was on direct thermal applications, which seemed to have the greatest chance for commercialization in the early 1980s. For the solar electric technologies, it was thought that broad-scale commercialization would not happen before the year 2000, although wind turbines might make a slightly earlier entry. The major decisions for solar electric technologies would be when to attempt large-scale demonstrations.

For each technology, key technical problems were identified; these were to be addressed and solved by the RD&D budget. In addition to the technical problems, the plan also addressed institutional problems, ranging from environmental concerns to the protection of sun rights, and from building codes to tax incentives. By this time, most of ERDA's solar managers were beginning to recognize that overcoming institutional barriers might be the toughest task before them. Initial congressional support toward overcoming the high "first cost" of solar systems was found in PL 93-383, the Housing and Community Development Act of 1974, which created a loan guarantee program to encourage lending institutions to accept the additional cost of installing a solar energy system when drawing up a mortgage. Because *ERDA-49* was advertised as a "National Plan," it included all the federal agencies, mandated or not, then participating in the program. Overlapping roles and responsibilities were rampant, but to a large degree beyond the control of ERDA's managers. For example, as can be seen in table 4.1, energy regulation and policy analysis alone would involve eight separate federal entities—a sure recipe for conflict. Fifty states and thousands of local governments would also have

Table 4.1
Federal areas of responsibility, solar energy (*ERDA-49*)

	ERDA	FEA	HUD	NASA	NSF[a]
Direct thermal applications					
Solar heating and cooling of buildings	×	×	×	×	×
Agricultural and process heating	×				×
Solar electric applications					
Wind energy conversion	×			×	×
Photovoltaic energy conversion	×			×	×
Solar thermal-electric conversion	×				×
Ocean thermal energy conversion	×				×
Fuels from biomass	×				×
Technology support and utilization					
Solar energy resource assessment	×		×	×	×
Solar energy research institute	×				
Technology utilization and information dissemination	×	×	×	×	×
Regulation of energy	×	×			
Energy policy analysis and development	×	×	×		×
Program definition	×	×	×	×	
Manpower and training	×				×

a. NSF responsibilies sharply curtailed after January 1976.

to be added to the list of involved stakeholders whose cooperation would be required if the unfamiliar solar technologies were to succeed. *ERDA-49* also addressed the patent and intellectual rights concerns discussed earlier. ERDA's patent policy was still seen as a potential "showstopper" when applied to the solar RD&D program. *ERDA-49* attempted to reduce industry's concern in this regard by highlighting in a separate section the many ways ERDA patent negotiators would find in favor of inventors and private businesses. It was hoped that the word would finally get around that ERDA was different from the AEC and would be a better business partner for the private sector.

ERDA-49 was the first true national solar energy research, development, and demonstration program. As we have seen, by the time this plan was formulated, the sweeping solar legislation passed by the 93rd Congress had been signed into law and the roles of the various federal agencies clarified. Coordination with other agencies, begun by NSF with IPTASE, was formalized with *ERDA-49*, and areas of responsibility were assigned. Table 4.1, from *ERDA-49*, shows how the program would be divided among the fifteen agencies involved, with specific mandated

STATE	DOD	DOC	USDA	GSA	USPS	HEW	FPC	FTC	TREASURY
	×	×		×	×	×			
			×						
	×		×						
	×								
	×								
	×								
	×		×						
	×	×	×						
×	×	×	×	×	×	×	×	×	
							×	×	×
		×						×	×
			×						
						×			

responsibilities assigned to nine: ERDA, NSF, HUD, NASA, DOD, GSA, FPC, DOC/NBS, and FEA.

Coordination between ERDA and FEA during this period had become especially contentious. FEA had assembled a small solar staff to carry out its new responsibilities, and they immediately began to lobby to increase their role. They were supported in the Congress by members still suspicious that the administration and ERDA had little desire to promote the use of solar energy. It was partly as a result of this lobbying that Congress enacted PL 94-385, the Energy Conservation and Production Act (ECPA), which assigned to FEA the lead role in solar commercialization. The act also required FEA to prepare a national plan for the commercialization of solar energy, a direct slap at ERDA, which felt that commercialization would be addressed, eventually, as the natural outgrowth of the program defined in *ERDA-49*. ERDA's reasoning was that until solar technologies had been successfully demonstrated in the marketplace, it was premature, and probably impossible, to put together a commercialization plan. One of the arguments of the Congress in favor of placing the commercialization role at FEA was that ERDA was an RD&D organization with no knowledge of the commercial world. ERDA's managers tended to look down on the new, inexperienced FEA staff but were forced to go along with the congressional mandate. Frank Zarb, having moved in the interim from OMB to FEA Administrator, testifying before the House Subcommittee on Energy Research, Development, and Demonstration in May 1975, stated: "We in FEA believe strongly that substantially expanded implementation must be made an integral part of the overall goal of a comprehensive national solar plan ... and ... it is time to emphasize the necessary next phase—accelerated utilization and widespread commercial application." Zarb's new approach differed markedly from his stance just a year earlier, when at OMB he turned down the NSF five-year, billion-dollar plan. Once again, the push for commercialization of solar energy by Congress was predicated on the assumption that solar heating and cooling was a technology ready to enter the marketplace. The same advocates who had clashed a year earlier with NSF during the debates that led up to passage of PL 93-409 continued to spread the premature notion that with a little government push, solar heating and cooling would be a major, early solution to our energy problems. As witness to this belief in the halls of Congress, fostered by overeager staffers, the following quote from a March 1975 letter from Chairman Olin Teague, of the House Committee on Science and Technology to Stever at NSF says it all: "During the hearings and reports leading to PL 93-409, Congress spoke especially to two major points that apparently were not recognized by NSF. One was that the technology for solar heating was available, and that *further research was not needed* [emphasis added]. The second was the urgency of making solar heating, and heating and cooling available to the public quickly."[5] The congressional advocates, as well as the new FEA staff, with little background in the status of the technology, truly believed in this position. The ERDA/NSF managers, with results starting to come in from early installations, held a much more conservative view and believed that continued RD&D was essential. The ERDA-FEA relationship never did become a smooth marriage. Fortunately, before their disagreements became too public, the election of 1976 intervened, and the new administration folded FEA and ERDA into the newly established Department of Energy. The FEA solar advocates thereby came under the management of the former ERDA senior staff.

The vision of a solar future for the country had taken another step forward with the release of *ERDA-49*. Whereas the earlier national solar plans carried only the endorsement of NSF, a small government agency,

ERDA-49 made bold to say, "ERDA believes that solar energy technology offers the potential for supplying as much as 25% of the Nation's energy needs from domestic resources by 2020." Solar heating and cooling, and direct thermal applications, whose projected contribution by 2020 was 20 quads, accounted for almost half the total solar contribution. Carrying the weight of the entire federal energy establishment, this was a statement that would have been unthinkable just twelve months earlier.

4.5.3 The National Program for Solar Heating and Cooling

By early 1975, thirty solar water heating and cooling demonstration projects were being designed, under construction, or operating across the country. Although most of these projects were funded by ERDA, NASA, or HUD, a few, such as the Institute of Gas Technology demonstration of dessicant air-conditioning units, were privately funded. This rapid increase in the demonstration of solar heating and cooling systems for both residential and commercial buildings was considered by many to indicate that such systems were ready for widespread market entry. This perception was shared by powerful members of Congress, a growing chorus of public advocacy groups, and some in the solar industry. ERDA managers were not as sanguine, especially for solar cooling applications. Early data returned from the school POCEs, which focused only on heating applications, began to show shortcomings in the current systems, ranging from installation to system operation and maintenance. A key mantra for solar energy enthusiasts was "life-cycle costing." One could justify the high initial installation costs by spreading the costs over a twenty-year operating life with fuel provided free of charge. Initial field experience, however, was beginning to demonstrate how difficult it would be to operate the existing solar heating systems for twenty years with reasonable maintenance costs.

Nevertheless, there was strong optimism that a growing RD&D program, supported by the government, would overcome the problems that were cropping up. Spurred on by the mandates in PL 93-409, the Solar Heating and Cooling Demonstration Act, and PL 93-473, the Solar Research, Developments, and Demonstration Act, ERDA made application of solar heating and cooling to residential and commercial buildings one of its first tasks. Although not in existence when PL 93-409 became law, ERDA inherited the responsibility to carry out its directive, "to provide for the demonstration within a three-year period of the practical use of solar heating technology and to provide for the development and demonstration within a five-year period of the practical use of combined heating and cooling technology." Because a year had already gone by since the act's passage, an interagency task force (an outgrowth of IPTASE now chaired by ERDA) quickly assembled an interim report (*ERDA-23*) and distributed it in March 1975.[6] The interim report solicited comments from all interested parties and promised that comments received by May 1 would be considered for both the Solar Energy Research, Development, and Demonstration plan due to Congress in June and the final Solar Heating and Cooling plan. This latter plan would also modify and incorporate the HUD/NASA plan for residential buildings sent to Congress at the end of 1974.

Because the National Plan for Solar Heating and Cooling was to be all things to all people, it started with three major elements:

1. Demonstrations for both commercial and residential applications;

2. Development in support of the demonstrations, initially utilizing available subsystems and components;

3. Research and Development to advance solar heating and cooling technology essential to the timely progress of the demonstrations and the eventual large-scale applications.

Based on comments received and further analysis by the interagency task force, the major elements were modified and strengthened for the final plan, *ERDA-23A*:[7]

1. Demonstrations of hot water heating, and space heating and combined heating and cooling for commercial and residential applications, initially using available systems installed in new buildings and retrofitted into existing buildings;

2. Development in support of the demonstrations, initially using available subsystems and components;

3. Research and advanced systems development to advance solar heating and cooling technology with particular emphasis on seeking improved solutions to the problems of retrofit systems.

The language in the final plan emphasized the problem of retrofit because it had been recognized by this time that if solar heating and cooling systems were to have a quick impact, the existing building stock would have to be targeted. In addition to the major RD&D elements above, the plan included elements on the collection and dissemination of information, on policy measures to achieve rapid and widespread utilization, and on program management, with a detailed analysis of each agency's responsibilities in the years ahead.

To cool overheated expectations, the plan addressed the question of technology status and the implications for life-cycle costing and early economic systems. The task force defined an economic system narrowly as a solar energy system for "new construction of certain heating-only applications, particularly in competition with electric resistance heating systems." Although this was a tightly constrained application, it was still considered to have a large target market because resistance heating was the space heating system of choice in much of the new housing market. The plan went on to say, "This, however, falls short of the program goal of realizing competitive solar systems for heating and cooling applications for a wide range of building types across the United States."

To place the question of solar energy systems cost in some context, in 1976 ERDA published *An Economic Analysis of Solar Water and Space Heating* (DSE-2322-1),[8] which stated that for new residential construction in "comparison with conventional energy [systems] costs, solar water heating and solar space heating installed at an equivalent cost of $20 per square foot of collector is competitive today against electrical resistance systems throughout most of the U.S." The report went on to say that if the system installation cost could be reduced to $15 per square foot of collector, solar energy systems would become competitive with oil-fired hot water heating systems and oil-fired and electrical heat pump space heating in many cities. Moreover, if system costs could be reduced to $10 per square foot of collector by 1980 through a combination of technology improvements and incentives, solar hot water and space heating would be economically competitive against all residential fuel types. This analysis was based on 1976 fuel costs for electricity, oil, and gas and installed solar energy system costs of $20 per square foot of collector.

Demonstration projects of available technology, carried out in conjunction with industry and users, were to be the chief mechanism used to develop a demand for solar energy systems. It was felt very strongly that creating a way to "kick the tires" of operating systems was the only way to get builders and buyers to accept this new technology. Yearly cycles of demonstrations were planned (up to five), with each succeeding cycle demonstrating more advanced technology in different applications. Three levels of programs were analyzed that would permit installation of between 350 and 2,000 residential units and 50 to 400 commercial units over a five-year period. Funding estimates for the three program levels ranged from $110 and $307 million for the five years.

The research element of this program covered a broad range, from materials and components, such as collectors and controls, to systems analysis, insolation and climatic data, and building design. This last item represented the first concerted effort in the government program to link active solar technologies with passive solar techniques and energy conservation. J. Douglas Balcomb of the Los Alamos National Laboratory had been an early champion of such a linkage, and through his efforts on the task force, the program finally included this important aspect of solar energy system design. However, it would be more than a year (FY 1977) before specific funds were included in the federal budget to support "passive" solar energy research.

Publication of *ERDA-23A* was the next step, after *ERDA-49*, toward clarifying management and oversight responsibilities for the various agencies involved in the solar heating and cooling program. As discussed earlier, the legislation passed by the 93rd Congress fragmented the primary responsibilities among nine agencies. In addition, almost all other federal agencies with buildings or facilities would eventually become involved. As with the solar RD&D program set forth in *ERDA-49* (see table 4.1), the interagency solar heating and cooling task force developed a complex management matrix (table 4.2), this time involving ninety separate activities needed to achieve the goal of moving solar heating and cooling from the research laboratory to the marketplace.

For those who wonder why government programs seem so disordered and complex, table 4.2 should supply an answer: we bureaucrats had made the program exceedingly complex in order to protect everyone's turf. Adding to the complexity, in part mandated by Congress, was the overall objective: solar heating and cooling commercialization. Government-led commercialization of a product was almost without precedent. One of the few examples that comes to mind is the synthetic rubber industry, which was born of wartime necessity, yet was both a wartime and peacetime success. Arguably, commercializing solar heating and cooling would be many times more difficult than developing a synthetic rubber industry in view of the enormous numbers of interfaces involved, from the government, to the manufacturer, to the builder/ installer, to the purchaser, to the local code board. All these actors and many more would influence the outcome of the effort. As the program evolved, FEA, which was supposed to lead the commercialization effort, was eventually assigned a lesser role, more in keeping with its capabilities.

Before its role was downgraded, however, FEA, following the mandate in PL 94-385, the Energy Conservation and Production Act, initiated studies in 1975 to plan for a Solar Energy Government Buildings Program (SEGBP). This was to be one element of a larger Federal Energy Management Program (FEMP) that FEA had defined in September 1975. The SEGBP studies identified a stock of approximately 450,000 government buildings or facilities located throughout the fifty states and the District of Columbia. Three agencies—DOD, GSA, and the VA— had responsibility for over 90 percent of all floor space in federal buildings, some 2.8 billion square feet and, along with the U.S. Postal Service, with its 36,000 buildings, were to be the prime targets for the installation of solar energy systems. The program was to be implemented with government funding appropriated either directly to the agencies targeted or to a central fund managed by FEA. In addition to saving fossil fuels, the SEGBP was designed to stimulate the solar industry through the provision of a large, mandated market whose needs would not be met for many years, by which time the nongovernment market would take over and keep the industry viable. FEA never formally initiated either FEMP or SEGBP, and both programs, the latter renamed the Solar Federal Buildings Program (SFBP), were inherited by the DOE when it was established in 1977. (For a thorough discussion of this aspect of the program, see volume 10 in this series.)

ERDA and HUD became the lead agencies for demonstrations and market development, while FEA was given the lead to develop regulations and incentives, a division of responsibilities that, from ERDA's perspective, was much more workable. NASA's responsibilities were also reduced in keeping with earlier ERDA decisions. In his first days at ERDA in early 1975, John Teem wrote to NASA's Harrison Schmitt to clarify agency roles in the implementation of PL 93-409, the Solar Heating and Cooling Demonstration Act, now that ERDA was in full gear. He stated that OMB had transferred energy RD&D functions to ERDA as provided for in the act and that ERDA would be responsible for technical

Table 4.2
Federal agency areas of responsibility for solar heating and cooling

Programe element task	ERDA	HUD	NASA	FEA	DOD	NSF	GSA	NBS	Other agencies
Residential demonstration	J	J	C	C	C		C	C	
A. Program design and management									
1. Demonstration design and location matrix	J	J			C/U				
2. Data requirements	C	L/C	C	C	C	C	C	C	C (NOAA)
3. Environmental impact statements	U	L		C					C (EPA)
4. Solar energy design and applications consultants	C	L/U							
5. Program management	C	L/U							
B. Establishment of performance criteria and standards									
1. Develop interim performance criteria		L	C		C	C		U	
2. Monitor demonstration for feedback	C	L	C	C	C			U	
3. Development of intermediate minimum property standards	J	J	U		C			U	
4. Develop definitive performance criteria	C	L	C		C	U	C	U	
5. Certification procedure	C	L	C		C			U	
6. Criteria for accreditation of testing labs	C	L	C		C			U	
C. Conduct of demonstrations									
1. Integrated project demonstration	J/U	J/U	C	C	C			C	
2. Subsystem and system selections	J/U	J/U	U		C			U	
3. System qualification and testing	C	L							
4. Site and site developer solicitation	C	L/U			C				
5. Design integration	C	L/U	C					C	
6. Construction	C	L/U							
7. Sales and occupancy	C	L/U			C/U				
8. Repair and maintenance of units		L/U			C/U				

The Buildup Years: 1974–1977

D. Data collection							
1. Collect and evaluate technical performance data	C/U	L	C/U	U		C/U	
2. Collect and evaluate market development and acceptance data		L/U		C/U	U	C/U	
3. Data for warranty program		L/U				C	
4. Data from nationwide survey and catalog preparation	L/U	C					
E. Market development							
1. Public barriers and constraints	C	L/U		C/U		C/U	
2. Economic factors	C	L/U		C/U	C	C/U	
3. Industry issues		L/U		C/U			
4. Market acceptance	C	L/U		C/U			
Commercial demonstration	L		C	C	C	C	C (USDA, USPS, DOI, HEW)
A. Program design and management	L/U				C		
1. Detailed design	L/U	C		C	C		
2. Technical design and data requirements	L/U	C		C	C	C	C (EPA)
3. Environmental impact assessment	L/U				C		C (EPA)
4. Utility options	L/U			C	C		
5. Potential market incentives	L/U			U	C		
B. Establishment of performance criteria and standards							
1. Develop interim performance criteria	L	C	C	C	C	U	C (USDA, USPS, DOI, HEW)
2. Monitor demonstrations for feedback	L	C	C	C	C	U	C (USDA, USPS, DOI, HEW)
3. Development of intermediate standards	L	C	U	C	U	C	U
4. Develop definitive performance criteria	L	C		C	C	U	C (USDA, USPS, DOI, HEW)
5. Certification procedure	L	C	C	C	C	U	
6. Criteria for accreditation of testing labs	L	C	C	C	C	U	

Table 4.2 (continued)

Programe element task	ERDA	HUD	NASA	FEA	DOD	NSF	GSA	NBS	Other agencies
Commercial demonstration (cont.)									
C. Conduct of demonstrations									
1. Integrate systems	L/U	C			C				
2. Subsystem and system selections	L/U	C	C		C				
3. Repair and maintenance of units	L/U	C			U		C		
4. Federal properties	L	C			U		U		U (USDA, VA, USPS, DOI, HEW)
D. Demonstration support and data collection									
1. Collect and evaluate technical performance data	L/U		U		C		C	U	C (USDA, USPS, DOI, HEW)
2. Collect demonstration data	L/U	C			C		C		C (USDA, USPS, DOI, HEW)
3. Prepare recommendations for achieving utilization	L/U			C/U					
4. Develop and publish manuals	L/U	C			U		U		U (USDA, USPS, DOI, HEW)
5. Data from nationwide survey and catalog preparation	L/U	U							
E. Market development									
1. Develop detailed financial procedures and recommendations supporting the use of solar energy systems	L/U	U		C/U	C/U		C/U		C/U (USDA, USPS, DOI, HEW)

The Buildup Years: 1974–1977

2. Develop standard tax appraisal and assessment procedures supporting the use of solar energy	L/U	U		C/U		C/U	C/U (USDA, USPS, DOI, HEW)
3. Develop consumer preference data	L/U	U		C/U		C/U	C/U (USDA, USPS, DOI, HEW)
4. Economic analysis techniques to predict savings potential of solar energy	L/U	U		C/U		C/U	C/U (USDA, USPS, DOI, HEW)
Research and development							
A. Research							
1. Components	L	C	C	C	C	C	
2. Applied materials research	L/U		U	C	C	U	
3. Building design implications	L/U	D	U	C	C	U	
4. System analysis	L/U	C	U	C	C	U	
5. Insolation and climatic data	L/U	C	U	C	D	U	
6. Advanced subsystem and system technology	L/U	C	U	C	C	U	U (NOAA)
B. Development for demonstration program	L	C		C	C		
1. System design and development	L/U		U				
2. Additional development of existing systems	L		U				
3. Systems integration of marketable subsystems	L		U				
4. Site data collection subsystem	L		U			C	
5. Central data processing system	L/U		U			C	
6. Additional development of existing subsystems	L		U				
7. Purchase of marketable subsystems (for B3 above)	L		U				

Table 4.2 (continued)

Programe element task	ERDA	HUD	NASA	FEA	DOD	NSF	GSA	NBS	Other agencies
Research and development (cont.)									
8. New subsystems for components development	L/U		U						
9. Demonstration engineering support	L		U					C/U	
10. Test and evaluation	L		U					C/U	
C. Advanced systems development									
1. System design and development	L/U		U		C	C			
2. System subsystems and components development	L/U		U		C	C			
3. System integration of marketable subsystems	L/U		U		C	C			
Collection and dissemination of information	L/U	U/C	U	U	C	C			
A. Demonstration information	L/U	U	U	U	U	U	U	U	U (all others)
1. Residential	J	J	U	U	U			U	
a. Central agency	L	U/C	U	U	U			U	U (NOAA)
b. Program activity	J	J	U	U	U			U	
c. Dissemination to building industry	C	L/U	U	U	U				
2. Commercial	L/U		U					U	U (NOAA, USDA, USPS, HEW)
a. Central agency	L/U		U					U	
b. Program activity	L/U		U					U	
c. Dissemination to building industry	C	L/U	U	U				U	
B. Supporting development information									
1. Central agency	L/U	U	U						
2. Program activity	L/U	U	U						
C. Research and advanced systems development information	L/U	U	U	U		U		U	U (all others doing R&D)

1. Central agency	L/U					U	
2. Program activity	L/U	U	U			U	U (all others doing R&D)

Additional policy measures for rapid, widespread utilization

A. Energy policy analysis and development

1. Overall energy policy as related to solar energy policy	C			L	C	U
2. Recommendations on constraint removal	C	C		L	C	U
3. Cost effectiveness of producer/user incentives	L	C		C	C	U

B. Program definition and analysis

1. Leverage required from different levels of demonstration	J	J		C	C	U
2. Number of units of initiate automation and development of industrial capability	J	J		C	C	U
3. Standards and criteria for near-term use of proven technologies	J	J	U	C	C	U
4. Enlarged program for solar equipped federal buildings	J	C	U	C	J	U, U (VA, all others planning construction)

C. Regulation and incentives

1. Utility related regulations	C	C		L	C	U
2. Conservation standards, new buildings	C	L		L	C	U
3. Building codes	C	L		C	C	U, U
4. Truth in energy labeling	L	C	U	C	C	U, U (OCA)
5. Environmental alternatives	J	J	U	U	C	U, U (EPA)
D. International activities	L	C	U	U	U	U, U (DOS, AID)

Legend: L = Lead responsibility   C = Consultation
J = Joint responsibility   U = "Utilizing the services of"

management of the solar demonstration program. He indicated that NASA's role would be primarily to procure and test systems and subsystems under ERDA direction. He also stated that a near-term action would be to form an interagency task force to review and modify the HUD/NASA demonstration plan submitted to Congress at the end of 1974. With NASA's greatly reduced overall role, this exchange renewed the agency's sense of exclusion from decision making on the solar heating and cooling program. In spite of this, management at Marshall Space Flight Center continued to support NASA's involvement in the demonstration program, and facilities and personnel resources expanded rapidly as the Center's space flight responsibilities were reduced. Over the next seven years, thousands of solar heating and cooling systems were installed under MSFC supervision on government buildings or at government facilities. However, although some operated successfully, the majority did not, and their impact upon widespread acceptance by the public of solar heating and cooling systems was largely negative.

4.6 Birth of the Solar Energy Research Institute

The birth pains of the Solar Energy Research Institute (SERI; known today as the National Renewable Energy Laboratory or NREL) offer a classical example of what happens when political expediency overtakes a basically sound idea.

SERI's right to exist was ordained by PL 93-473, the Solar Energy Research, Development and Demonstration Act of 1974, signed into law in October 1974. Its implementation was left to the soon-to-be-established ERDA. The law stated:

> There is established a Solar Energy Research Institute which shall perform such research, development and related functions as the Chairman may determine to be necessary or appropriate in connection with the Project's activities under this act....

(When ERDA was created, the "chairman" named in the Act became the administrator of ERDA and the institute's programs became ERDA's programs.) Because PL 93-473 did not spell out the mission and structure of SERI, ERDA had considerable latitude in implementation. John Teem, the Assistant Administrator for Solar, Geothermal and Advanced Energy

Systems, was assigned responsibility for developing the plan to bring SERI into existence. While awaiting Senate confirmation of his nomination, Teem, as one of his first official acts, wrote to President Philip Handler of the National Academy of Sciences (NAS) on February 21, 1975, asking that NAS and the National Academy of Engineering explore in depth the role and desirable characteristics of such an institute. At the same time, the MITRE Corporation was awarded a contract to assemble industry opinions on the formation of SERI, and later, on June 3, 1975, ERDA published a notice in the *Federal Register* soliciting opinions from the public on SERI's formation.

At ERDA, a three-phase plan was developed that would define the role, mission, and type of management organization needed by SERI and culminate with the selection of a manager and site. Robert P. McGee was appointed manager of the Solar Institute Project Office (SIPO) and given a small staff to oversee the plan. The first two phases were to lay the ground for the release of an open solicitation for selection of SERI's management and site. Robert Seamans would be the source selection official for what would ultimately be a highly charged, political decision.

By October 1975, all the activities began to coalesce. The NAS committee, chaired by Richard L. Garwin of the IBM Corporation, published its report.[9] Among its many recommendations, three were especially cogent: (1) that there be a single SERI with a number of small, supporting field stations, (2) that SERI be operated under contract to ERDA (similar to the arrangement for the former AEC laboratories), and (3) that SERI cover activities from basic research to manufacturing research and technology transfer. The NAS committee recommended that over the course of three years SERI should grow to a staff of 630 professional people (within a total laboratory population of 1,430), requiring an annual budget of $48 million, in 1975 dollars. The committee also provided site criteria for the location of the main institute as well as the field stations in the form of a checklist, which made large parts of the United States potentially eligible for selection. All of the above attracted the attention of many communities just then struggling to recover from the recession that had been brought on by high oil prices. Whoever landed SERI would have a plum indeed since no other similar federal projects were on the horizon.

Like virtually all groups assessing the status of solar technologies, the NAS committee assumed that solar heating and cooling technology was "well understood and demonstrated in a variety of applications"—a surprising perception in view of the short period (two years) in which the small NSF R&D program had been under way, and the small number of projects operating or in development. NSF's public relations efforts were evidently more successful than anticipated. The committee did note that the technology was not competitive with systems based on oil and natural gas; it recommended additional basic R&D, to include materials development and testing, and called for a strong effort in system economics, market studies, and the development of performance standards. Staff requirements for these efforts were estimated to be 62 professionals. Interestingly, the committee made no mention of the need to study or advance passive solar techniques and technology, an omission that reflected the overall government-funded RD&D program, which had until then almost completely ignored this aspect of solar technology.

The MITRE solar energy report that surveyed industry and regulated public utilities, also published at this time, contained no recommendations and was strictly a compilation of opinions received.[10] The industry population polled ranged from large companies such as Exxon and Grumman to small companies such as InterTechnology Corporation (ITC) and Solaron Corporation. The only criteria for being selected to participate in the study were that an organization currently participate or plan to produce solar energy equipment, engineer or install equipment, or operate solar energy systems. Although there were some dissenting opinions on the need for an institution like SERI, three principal functions for such an institute emerged as a result of the MITRE poll: (1) to disseminate information and data resulting from government programs, (2) to undertake basic research and systems analysis, and (3) to develop test and evaluation procedures and performance standards. Industry opinion was thus in agreement with the NAS recommendations on the mission of a solar energy research institute.

In response to ERDA's *Federal Register* request for opinions from the public, hundred of letters were received, among which were many from congressmen and senators touting the attractiveness and availability of sites within their states or districts, presaging the intense competition that would soon develop.

Within the ERDA bureaucracy, reaction to the idea of the institute was decidedly mixed. Many solar energy managers saw it as a direct threat to both their budgets and their authority. This attitude was to persist throughout SERI's early history.

In an attempt to consolidate the opinions received by ERDA, SIPO established two study groups to document site criteria and alternative missions for a solar institute of differing sizes and responsibilities. The second study group report, *Alternatives and Recommendations Related to the Mission, Role, and Management of the Solar Energy Research Institute*, proved to be highly controversial. Prepared in draft form in December 1975, the study developed a rationale for four different types of institutes. Briefings with Bob Seamans and John Teem narrowed SERI's mission only slightly from the broad mission recommended by the NAS. Because OMB had to be in agreement before a solicitation was released, meetings were held to receive their blessing. It was at this point that OMB and the White House Domestic Council expressed strong reservations as to SERI's future role. One of the major sticking points was SERI's proposed $50 million/year budget. In-house studies had recommended that SERI's budget be about 10 percent of the annual ERDA solar budget, which translated to an annual ERDA solar budget of $500 million, far larger than what OMB had under consideration in 1975. OMB was also concerned, correctly, that the siting of a solar institute with major responsibilities and large budget authority would become a political football. Eventually, a compromise was reached scaling SERI back slightly, but with the understanding that no award would be made until the results of the 1976 election were known and the next administration was in place.

With the ground rules agreed upon by all participants, it was now time to start the procurement clock. Seamans was determined that procurement would follow established ERDA rules laid out in the *ERDA Source Evaluation and Selection Handbook* to avoid postselection problems. One of the first steps was to select a chairman for the Source Evaluation Board (SEB), and on March 2, 1976, he selected Assistant Director Raymond Fields, Office of Thermal Applications, Division of Solar Energy. Fields had been one of the original NSF solar managers brought into ERDA at its formation and had been involved in many of the early solar heating and cooling demonstrations. Seamans also appointed five other members from various other ERDA organizations, including the Office of the General Counsel and Division of Procurement.

It was SEB's responsibility to oversee the development of the Request for Proposals (RFP) and ensure that its content and language would help them make a reasoned recommendation to the designated Source Selection Official, Bob Seamans. With most of the groundwork already laid by the two SIPO study groups, the RFP was released on March 15, 1976. Its closing date of July 15, 1976, allowed for an unusually generous response time of 120 days. The SEB released the RFP with breath held, hoping they would not have to deal with fifty proposals, one from each state, at the closing date.

Twenty proposals were received, each accompanied by a strong congressional statement of support. One proposal was considered nonresponsive, and nineteen went forward for evaluation. The evaluation and selection process followed established procedures, and eventually, under tight secrecy, a winner was selected but not announced, pending the review of the next administration. But as often happens in Washington, the selected winner's name and site soon became known.

As agreed earlier with the OMB, on March 23, 1977, with a new administration in power, acting ERDA Administrator Robert W. Fri and ERDA staff met with James R. Schlesinger at the White House to review the status of the SERI competition and the selection of the Midwest Research Institute (MRI), with Golden, Colorado, as the site. Lobbying by members of Congress had been underway for some time in an attempt to sway the Carter Administration in its award, and rumor had it that Speaker Tip O'Neill of Massachusetts was at the forefront of these efforts that included Senator Hubert Humphrey (D-MN). During the transition period before a newly elected administration is inaugurated, it is common practice for agencies and departments to develop lists of upcoming events or problems that will face the new administration. Near the top of ERDA's list was the decision on SERI. In the exchange that took place between ERDA and Schlesinger's staff, the response that came back to ERDA in December 1976 was: "The [transition] paper does not squarely propose an alternative to the selection of a single solar energy R&D institute. We wish a specific alternative to the proposal which would create a network of institutions throughout the nation." This was a clear signal to ERDA that the lobbying for multiple SERIs was having an effect. ERDA replied that a great deal of dialogue had already taken place to arrive at the single-SERI approach and that "the reaction of the proposers, the Congress and the public would be extreme if the SERI approach were altered at this late date." As a result of the Schlesinger meeting it was agreed that MRI and the Golden site would be announced as the winner for the "national" SERI award but, in a last-second bow to political pressures, that there would be three additional sites, in the New England, North Central, and Southeastern regions. Each region would include the contiguous states, which would join the regional SERI on a voluntary basis. The specific location of each regional institute would be determined later. This capitulation to political expediency, which ran counter to the NAS recommendations and, one could argue, to the language of the act establishing SERI, would eventually lead to programmatic, budgetary, and political problems culminating in the special congressional hearing of October 1979.

By the end of April 1977, the Golden SERI was already operating informally, under the leadership of Paul Rappaport; an exploratory meeting was held in Denver with the interested regional coalitions to explain the process that would be followed. In early May all the regions, with the exception of New England, which got an early jump and planning grant through the intervention of Speaker O'Neill, met separately to get their proposals in order and to apply for planning grants. Some states, arguing for more than three regional SERIs, attempted to coerce the administration to accept additional, smaller coalitions. These attempts were rejected, but not without creating long-lasting bad feelings. Moreover, you may have already noticed a large regional void, with no regional center planned in the West or Southwest, two important areas for solar development. The reason for this omission was that when the California SERI proposal, which included New Mexico, was turned down, the proposal team publicly criticized the selection process and did not lobby actively for a regional center. Eventually peace was made, and despite a late start, a western regional team also joined the hunt.

Each region was given considerable latitude in determining the form and technological emphasis of its program. It was hoped that with guidance from Washington and the "national" SERI, each region could identify technologies and problems unique to the region, to avoid unnecessary overlap and competition for funds. Eventually four Regional Solar Energy Centers (RSECS) were established, headquartered in Massachusetts, Georgia, Minnesota, and Oregon, with tenuous ties to the "national" SERI in Colorado. (For additional details on the regional solar energy centers or RSECs, see chapters 5 and 6, this volume.)

4.7 John Teem's Resignation

As the first confirmed Assistant Administrator for Solar, Geothermal, and Advanced Energy Systems, John Teem assumed responsibility for uniting under one office the most disparate functions of the newly formed ERDA. The "Advanced Energy Systems" part of his title included programs in magnetic confinement fusion, high-energy physics, and basic energy sciences, which entailed research in virtually all scientific disciplines. The solar and geothermal programs included not only those originated in NSF and AEC but a burgeoning number of programs coordinated with other government agencies and cabinet-level departments. Although they accounted for less than half of the funds under his direction, the solar programs soon began to occupy most of his time.

The contentious nature of the solar programs guaranteed that Teem would never be able to satisfy all the protagonists. As 1975 came to a close, and after less than one year on the job, he found himself more and more at odds with the direction and budgets he was receiving from the administration and OMB. In particular, he felt that the scope and pace of the Solar Heating and Cooling of Buildings Program could be increased to be more in line with congressional recommendations. He was also concerned with the difficulties encountered in moving ahead with the establishment of SERI. The administration and OMB were clearly dragging their feet and erecting road blocks. He correctly foresaw that developing SERI's mission and budget as well as selecting its site would lead to increased friction with the Congress and eventually lead to hearings. These matters, as well as a concern with the competition with NSF on basic research, led him to resign at the end of January 1976. In an interview with the *Wall Street Journal*, reported on February 26, 1976, Teem reiterated these themes and indicated that, although his official resignation letter cited only "personal reasons" for leaving, he believed his proposals would have resulted in more government stimulation of private sector commercialization of solar energy than the White House and the OMB wanted. His resignation confirmed the suspicions of solar supporters, especially those in Congress, that ERDA and the administration were not going to be solar energy champions.

In spite of Teem's misgivings and the loss of his leadership, 1976 proved to be a very successful year for solar programs. Robert S. Hirsch was quickly nominated and confirmed to succeed Teem. Almost all solar energy RD&D had now been transferred from NSF to ERDA. Support on Capitol Hill for NSF's RANN program to continue energy research had essentially vanished, and the Congress returned an appropriation to NSF below the already reduced administration request. Any potential competition for funds and mixed signals as to which agency would sponsor solar energy research had been eliminated. Solar budgets passed by the Congress for ERDA were keeping pace with the aggressive solar program called for in *ERDA-49* and *ERDA-23A*.

The vast laboratory system ERDA had inherited from the AEC lobbied to become part of the solar programs. With their missions tied almost exclusively to DOD weapons programs, many researchers at the laboratories saw ERDA's multifaceted programs as an opportunity to come out from behind the cloak of secrecy and participate in highly visible (and reportable) research. The solar program managers, in particular, were eager to tap into this reservoir of talent and facilities to further their research goals. With the Cold War in full swing, however, the management of the laboratories (composed of former AEC people) was reluctant to dilute defense programs by diverting resources to new missions. The sometimes contentious debate over the role of the national laboratories was eventually resolved by General Alfred D. Starbird, Assistant Administrator for National Security Programs, who agreed that up to 10 percent of the work at the laboratories could be in support of non-defense-related activities. Because the laboratories were carrying out billions of dollars worth of research every year, 10 percent represented a potentially enormous increase in the efforts that could be assigned to these premier research institutions. The solar program managers quickly took advantage of this opening and expanded programs at the Sandia, Los Alamos, and Lawrence Berkeley laboratories.

By the end of 1975, over 300 solar-heated buildings had been completed or were under construction in the United States, the vast majority funded by the NSF/ERDA solar programs, a tenfold increase in less than twelve months. As required by PL 93-577, the Federal Nonnuclear Energy Research and Development Act of 1974, ERDA revised its first national plan for energy RD&D, *ERDA-48* (issued in 1975), and published *ERDA 76-1*, with the same title.[11] The major change in emphasis from *ERDA-48* was in singling out conservation technologies to be included in the highest-priority programs. Funding levels in the president's budget request for conservation increased 64 percent over the previous year. Solar energy

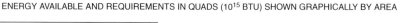
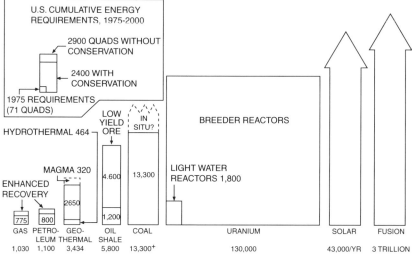

Figure 4.2
Recoverable domestic energy resources.

continued to be a high-priority program, with large budget increases and a growing comfort level that solar technologies could make a significant contribution. *ERDA 76-1* projected a 2-quad energy savings if only 10 percent of the existing building stock were solar-equipped by the year 2000—a percentage that was considered attainable. A new energy graphic (figure 4.2) found its way into ERDA briefings and was used frequently by senior ERDA management as they traveled the country advocating aggressive energy RD&D programs.

Solar energy was shown as a potential major contributor, sharing the future with two former AEC pet programs—the breeder reactor and fusion energy—and dwarfing conventional sources of coal, gas, and oil. *ERDA 76-1* acknowledged that many technical problems needed to be solved before solar's full potential could be realized and tempered any enthusiasm by stating that "solar energy technologies and their applications will require varying degrees of further development before they can become economically viable."

Whereas *ERDA-48* did not foresee any special constraints to implementation of solar energy technologies, *ERDA 76-1* identified several, centered on high initial system costs. For solar heating and cooling, these constraints echoed the concerns of *ERDA-23A* and resulted in the identification of a six-element "Action Program":

1. Residential demonstrations;
2. Commercial demonstrations;
3. Development in support of demonstrations;
4. Research and advanced systems development;
5. Collection and dissemination of information;
6. Additional policy measures to achieve rapid and widespread utilization.

The program had taken on an added "kick the tires" flavor, with demonstrations of heating and cooling technologies as the centerpiece. ERDA recognized that for the demonstrations to be successful, industry involvement was essential. In particular, heavy emphasis was to be placed on the involvement of small business for the manufacture, installation, and servicing of solar energy systems. The management roles of ERDA and HUD in the demonstrations had been clarified, and several POAs and PONs had been released. At the end of 1975, ERDA published *ERDA-75*: "Catalog on Solar Energy Heating and Cooling Products." This catalog offered the first comprehensive listing of the more than 300 companies manufacturing components or systems. It was expected that numerous responses would result from the POAs and PONs to demonstrate solar heating and cooling systems, the first two elements of the "Action Program," and this is in fact what happened. The programs were oversubscribed, with many more proposals received than could be funded. Similarly, responses to participate in the other four elements engendered overwhelming interest resulting in dozens of new contracts.

In April 1976, Hirsch announced a series of awards totaling $7.5 million for 34 nonresidential solar heating and cooling demonstration projects in 22 states plus the Virgin Islands. The projects included 10 office buildings, 4 schools, 3 hotels or motels, 2 fire stations, 1 hospital, 1 laboratory, and 1 library. The size of the projects ranged from over $1 million for a solar hot water system at Trinity University in San Antonio, Texas, to less than $5,000 for a space and hot water system at the Lake Valley firehouse in Lake Tahoe, California. The awards went to both large and small companies and augmented 31 projects already accepted from unsolicited proposals received earlier at NSF.

At the same time, ERDA provided DOD with $900,000 to supplement projects proposed at Randolph Air Force Base in Texas and Kirtland Air Force Base in New Mexico to install solar heating and cooling systems in the large base exchange shopping centers.

The Industrial Process Heat Program also made its first major awards in June 1976. Out of the 43 proposals received, 4 design contracts were selected: a solar hot water system for curing concrete blocks, a system to provide hot water to wash cans in a canning factory, a system for textile dyeing, and another hot water system for a commercial laundry. All of these applications were in keeping with the targeted industrial applications that used relatively low temperature (below 350°F) process heat or hot water, estimated to be 16 percent of total U.S. energy demand. In addition, Battelle Laboratories of Columbus, Ohio, and InterTechnology Corporation (ITC) of Warrenton, Virginia, were given large awards to identify other industrial processes that easily lent themselves to the application of solar systems.

Another highlight of 1976 was the Bicentennial Solar and Energy Conservation Exhibit held on the Mall in Washington, D.C. With a flood of visitors expected in Washington for the bicentennial celebration, ERDA, along with FEA and HUD, decided to showcase recent advances in solar energy and conservation. A full-scale model house was built to reveal, in cutaway views, the latest energy-conserving methods and materials, including a roof-mounted solar collector system. Because the exhibit opened in mid-May and was scheduled to operate through most of the summer, it included a visitor rest area cooled by a second solar-powered system. Computer terminals resembling phone booths were also built, in which visitors could enter their telephone area codes and receive printouts of information to be used in home energy conservation and planning for the design of a solar energy system sized for the local area. The exhibit proved very popular. By the end of 1976, despite continuing serious personnel shortages at ERDA, all elements of the "Action Program" were moving ahead and receiving high visibility as the ribbon cuttings on new buildings proliferated. The program had far surpassed NSF's magic number of 535.

4.8 Barriers and Incentives to the Use of Solar Energy

Barriers to the widespread utilization of solar energy systems, in addition to the R&D needed to reduce cost and improve reliability, were first outlined in 1973 in the Subpanel IX report. Specific incentives to overcome these barriers were not called out in that report, however. A year later, the Solar Energy Task Force report for Project Independence called for a systematic effort to overcome institutional and regulatory constraints, and specific funds to address these problems were requested in the National Solar Energy Program submitted to Congress at the end of 1974. A few small-scale studies were begun to address the problems within the limited budgets available.

With the advent of ERDA and expanding budgets, a Barriers and Incentives Branch directed by Roger H. Bezdek was established in John Teem's office. The need for this specific office became evident after the publication of *ERDA-23A* and after the tasks assigned to the various agencies had been clarified. ERDA, HUD, and FEA were given responsibility for undertaking studies in market development that included analysis of incentives. ERDA management was anxious to take the lead in developing an overall strategy in this increasingly important area. Among the branch's first tasks was a study of the incentives that had been accorded conventional energy sources of oil and gas, coal, and nuclear energy. The underlying rationale for this study was to quantify these incentives and then make the case for parity of treatment for renewables. The results of this initial study proved to be enormously important to the renewables programs. Incentives in the form of direct subsidies, regulation, tax incentives, market support, R&D, demonstration programs, procurement mandates, technology transfer, information generation, and other government actions had subsidized and entrenched the conventional energy sources now competing with renewables. Dating back in some cases to the 1920s, these subsidies totaled, through FY 1977, $211 billion (in constant 1977 dollars).[12]

Armed with this information, the solar energy programs began a low-profile lobbying campaign to sway the Ford administration and Congress to accord the same types of subsidies to solar energy. Starting in 1976, a series of analyses was undertaken to determine what types of government actions would be most beneficial in accelerating the market penetration of solar heating and cooling systems and agricultural and industrial process heat, the technologies with the greatest potential to make a near-term impact. The incentives analyzed were varied: 20 percent investment tax credits, 5 percent (low-interest) loans, taxes on energy use for conventional energy forms, a billion-dollar federally supported demonstration program, oil and gas deregulation, and various other regulatory actions.

It was believed that any or all of these would significantly stimulate the newly developing solar energy industry.

The lobbying was effective. In April 1976, speaking before the Senate Joint Economic Committee, Representative Harold Runnels (D-NM) stated that solar energy "continues to get a second-class treatment by Congress and the Administration" and that, to his knowledge, "there has been no legislation proposed, to date, that will provide the same kind of incentives to the solar energy industry that were provided for decades to the fossil fuel industry." Legislation was soon introduced to accomplish this goal.

These analyses, made public in the spring of 1977, proved timely. The Carter administration, beginning its Domestic Policy Review of Solar Energy, utilized the numbers that came out of the analyses to justify its bullish estimate that solar energy could provide 20 percent of the nation's energy needs by the turn of the century. Congress, which had already begun to debate the use of incentives to hasten the introduction of solar energy systems, also made use of these ERDA analyses. The 95th Congress, like the 93rd, established an activist agenda supporting renewable energy and energy conservation. Significant legislation was passed to provide incentives for the introduction of conservation and new energy sources, including PL 95-617, the Public Utilities Regulatory Policy Act (PURPA), which required utilities to buy and sell power from interruptible sources at fair rates. PURPA began to change the way the established energy infrastructure looked at the utilization of new energy sources. Coupled with the growing public skepticism on the siting and building of new power plants, especially nuclear plants, PURPA began to level the playing field for solar energy.

4.9 Major ERDA Energy Studies

After two years of struggling to put all of its RD&D programs and funding within some logical framework, ERDA management decided in early 1977 to initiate detailed studies that would capitalize on the latest knowledge of technology status and the changing marketplace.

4.9.1 The Market-Oriented Program Planning Study (MOPPS)

Perhaps the most controversial energy study undertaken by the government during this period was the Market-Oriented Program Planning

Study (MOPPS). Shortly after Bob Seamans resigned from ERDA in January 1977, Acting Administrator Bob W. Fri requested the MOPPS. With the incoming Carter administration committed to new priorities in the energy field, Fri felt it necessary to have in hand an analysis that would allow him to make rational decisions for allocating RD&D funds among the many competing technologies, taking into account such difficult projections as supply/demand, environmental constraints, and technological advances.

The MOPPS was to take a new approach toward developing RD&D funding strategy by establishing technology goals related to market sectors and the anticipated contribution each technology would make in each sector. Projections would be made based on date of entry, rate of penetration, comparable economics, and possible environmental/institutional barriers. All of this was to be done using a common set of assumptions and a common database, an approach never before applied rigorously. Fri appointed Philip C. White, Assistant Administrator for Fossil Energy, to chair the study. Because of the press of White's official duties, Fri soon added Roger LeGassie as cochair. Harry R. Johnson, who headed White's planning staff, was named executive director of the study, thus giving it a heavy fossil fuel energy bias. Originally, the MOPPS was to exclude technologies aimed at long-term inexhaustible resources, identified as solar electric, fusion, and the breeder reactor; a second study, the Inexhaustible Energy Resources Study (IERS), was to handle these technologies. These ground rules were soon suspended, though, and the MOPPS included all potential supply technologies.

By April 1977, preliminary MOPPS results were available and briefings for ERDA management began. These briefings were a prelude to the briefing of President Carter's new energy czar, James Schlesinger, and his staff, which was scheduled to take place before President Carter delivered an energy message on national television on April 20. The initial briefing created major disagreements within ERDA management because it indicated that great quantities of relatively low priced natural gas would be available if the regulated price of natural gas was permitted to climb to competitive energy prices. This premise was contrary to the belief held by some that we were running out of natural gas. With large gas supplies available at increasing prices, few new technologies would be able to compete in the near term. A second, parallel study, chaired by Martin Adams, did not confirm this rosy a picture of the natural gas supply, predicting that the availability of large quantities of gas would come at a later date; but the damage had been done, and it was too late to put the MOPPS analysis back in Pandora's box.

Subsequent Senate hearings resulted in much finger pointing and name calling and talk of cover-ups reminiscent of the acrimonious congressional hearings on the Subpanel IX report three years earlier, as special interests on both sides of the debate attempted to justify their positions. In the end, because of its controversial content, the draft MOPPS report, circulated in December 1977, was recalled and never formally released.[13] From the standpoint of solar energy, the MOPPS was very negative concerning the penetration solar heating and cooling would make in the residential and commercial sectors and predicted that these technologies would save or substitute for only .06 quad of energy by 1985 and .24 quad by the year 2000. Solar electric technologies would make an even smaller impact.[14] The new administration and Congress chose to ignore the MOPPS' projections and instead rapidly increased the solar energy budgets. In retrospect, however, the MOPPS projections for solar energy contributions by 1985 were not far off the mark, although the underlying determinants were quite different than those projected by MOPPS.

4.9.2 The Inexhaustible Energy Resources Study (IERS)

As mentioned earlier, the IERS was to be undertaken simultaneously with the MOPPS. Because of the press of activities associated with the transition to the Carter administration, Bob Fri did not formally initiate the study until April 1977. James S. Kane was named study coordinator, and Bennett Miller executive director.

For those working in the solar trenches, the unfolding MOPPS projections were viewed with great alarm. They were rumored to be the first step toward a selection of energy RD&D program winners and losers. Because members of the solar energy program staff participated in the MOPPS, and were familiar with the assumptions being used, they believed that the MOPPS conclusions would unfairly penalize new technology and, in particular, solar technologies. The IERS was therefore looked upon as a way to refute some of the MOPPS conclusions. Like the MOPPS, the IERS was expanded in scope; instead of confining its solar energy analyses to solar electric technologies, it eventually included solar heating and cooling and agricultural and industrial process heat.

The IERS, like the MOPPS, was fated never to be officially released. Instead, there was one public meeting to discuss preliminary results in July 1977, and a thick file of analyses was developed that was, to some extent, utilized in later studies undertaken during the Carter administration. From a solar technology perspective, the IERS was much more bullish than the MOPPS on the projected impact of solar energy. Using an approach that emphasized the question "How soon will inexhaustible energy sources be needed?," the study defined a "window of deployment." As in all the preceding studies, IERS chose to analyze two different scenarios, called "reference" and "stressed-case." The "stressed-case" scenario envisioned events that would dictate an earlier introduction of alternative energy resources. Like the MOPPS, the IERS subjected each inexhaustible energy source to identical assumptions, modeling, and evaluation criteria.

Based on the IERS's elaborate set of assumptions, the "window of deployment" of solar technologies for the two different scenarios was from 1997 to 2015, with solar heat technologies and energy from biomass being viewed as more advanced and having an impact by 1990. Although providing a slightly different time frame from the MOPPS—1990 versus 1985—the IERS projected an energy substitution of 1.05 quads in 1990, as against the MOPPS figure of 0.06 quad in 1985. (Table 4.3 should help the reader in tracking the history of solar energy projections from the FCST Solar Energy Panel through 1977.)

Even though the IERS was not brought to completion and did not have the impact of the MOPPS, it did help provide some legitimacy to the new administration's Domestic Policy Review of Solar Energy. In the minds of ERDA's solar managers, soon-to-be DOE managers, it had blunted the MOPPS ax.

4.10 Budgets and Budgets

The rancor that had developed between Congress and the White House during the 1970s over solar energy budgets was fueled, in part, by what today would be considered "leaks," but which, in the 1970s, were perfectly acceptable early releases of working-level budget numbers. The long, tortuous process leading up to the president's signing an appropriation bill has many twists and turns. Authorization and appropriation

Table 4.3
Projections of solar energy impact based on government studies (impact in quads, excluding hydroelectric)

Study title	Year of study	1985	1990	1995	2000	2010	2020
NSF/NASA Solar Energy Panel	1972	0.3+			11.8		109
The Nation's Energy Future Subpanel IX	1974	2.0+			17.44		
Project Independenct	1974						
"Business as Usual" scenario		1.2	3.7	7.0	16.0		
"Accelerated" scenario		1.8	7.0	16.0	44.0		
A National Plan for Energy RD&D (*ERDA-48*)	1975	small			11.0–15.5		
National Solar Energy RD&D (*ERDA-49*)	1975	0.8			10.0	2.0	45.0
Market Oriented Program Planning (MOPPS)	1977	0.06			1.37		
Inexhaustible Energy Resources Study Base case 2	1977		1.05			5.8	29.5
President's Council on Environmental Quality	1978				24.5		45.0
Domestic Policy Review of Solar Energy	1979				6.0–14.2		
Estimated range of total U.S. energy usage		95–120			100–180		105–300

committee oversight hearings are just one step in the long chain, with the president's budget request constituting the first, official public step. But even before the president submits his budget, protracted debates go on behind the scenes between the agencies and OMB.

ERDA's budget preparation carried with it an AEC heritage. Beginning as early as 1957, AEC had provided budget detail to the Congressional Joint Committee on Atomic Energy (JCAE) that normally was considered "privileged information" by other agencies and therefore not usually provided to Congress. Specifically, AEC had provided internal agency working papers or documentation of requests to the Bureau of the Budget (BOB; later the Office of Management and Budget or OMB), to justify budgets that BOB might or might not approve. These communications, initially requested by the JCAE, eventually achieved a life of their own and became known as "Holifield Tables" after Senator Chet Holifield (D-CA), the chairman of the JCAE. These tables gave the committees means to second-guess BOB/OMB and change the president's request if it suited their purpose.

One can easily see the political games that could be played. Because agency budgets, to some degree in response to OMB targets, were bottom-up aggregates starting at the division level, division budgets usually represented wish lists with large increases over the previous year. After all, in Washington, a good manager was a manager whose program was growing. This could lead to some interesting and imaginative requests, with managers often finding themselves in the position of making budget requests for future fiscal years before they knew what their actual budgets would be for the current and following years. As a rule, three budgets were on the desk of every project manager: the operating budget for the current year, the budget being considered by Congress for the next fiscal year, and the budget under development for submission to the administration for two fiscal years in the future. Because solar budgets were growing rapidly at this time, a manager would not be too far out of line to project rapidly increasing growth.

And so it came to pass, in the spring of 1976, that ERDA management was caught up in the congressional enthusiasm, not shared by the OMB, for large solar RD&D budgets. The president's FY 1977 request, sacrosanct to all agencies and to be defended to the last breath during agency testimony before Congress, was $162.5 million for solar energy. The authorization and appropriations committees, privy to the Holifield

Tables, knew that the ERDA Solar Energy Division had originally requested $300 million. Although this number had been massaged by John Teem and reduced to $255 million before it left ERDA for OMB, it was more than double the previous year's request, including the unusual transition quarter that marked FY 1976. Under direct questioning in the hearings that followed, Teem's successor, Robert Hirsch, was allowed to put these larger numbers in the record and to agree that he could intelligently spend more than the $162.5 million in the president's request. With any other agency, a presidential appointee making such a statement would have been destined for early retirement. With ERDA, it was an accepted part of the game. Congress in its wisdom passed a $290.4 million solar budget for FY 1977, the president signed it, and solar energy RD&D was off to the races. Included in the $290.4 million was $92.6 million for Heating and Cooling and for Agriculture and Process Heat, plus some funds for plant and capital equipment.

Figure 4.3 plots a history of the budget battle between Congress and the White House from FY 1974 through FY 1979. Each year, Congress

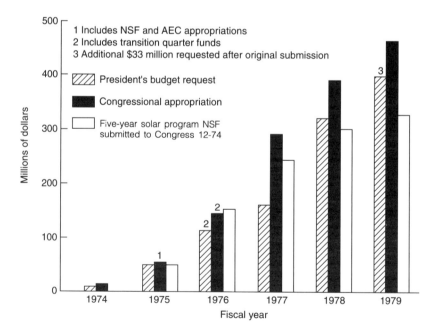

Figure 4.3
Solar energy programs funding history, 1974–1979 (in current year dollars).

added funds, sometimes significantly, to the president's request. The table also shows how the appropriations compared to the five-year plan (starting in FY 1975) submitted by NSF to Congress at the end of 1974. This billion-dollar-plus plan, discussed earlier, had funding levels very similar to those submitted for Project Independence and subsequent studies such as MOPPS, IERS, *ERDA-49*, and *ERDA-23*. As can be seen, by FY 1977, Congress began to exceed these totals, and by the following year, administration requests also surpassed the funding runouts originated five years earlier. For the ERDA and DOE managers who had persevered through these contentious years, a sense of vindication was palpable.

4.11 Establishment of the Department of Energy

With the end of the OPEC oil embargo, the energy crisis of 1973–74 receded into the background. Gasoline was plentiful, although more expensive, and most sectors of the economy had adjusted to the overall higher energy prices. In spite of the general loss of public concern over energy matters, the Ford administration continued to search for ways to strengthen the executive branch's ability to deal with energy policy and future energy emergencies. In both 1975 and 1976, President Ford proposed to Congress a number of changes including the establishment of an Energy Independence Authority that would coordinate the drive toward energy independence first proposed by President Nixon. As a result of the give-and-take between the Ford administration and Congress, a new law was enacted: PL 94-385, the Energy Conservation and Production Act (ECPA), that had a direct bearing on the future growth of solar energy programs. Finally, as one of the last actions of his administration, President Ford submitted to Congress an energy reorganization proposal (not acted upon) that included creation of a cabinet-level Department of Energy.

After the election of November 1976, a new emphasis was placed on the "energy problem." This emphasis was reinforced within the new Carter administration by a bitterly cold winter, which resulted in energy shortages and, once again, outcries from the public sector to solve the problem. President Carter went so far as to call the resolution of the country's energy problems the "moral equivalent to war." Intelligence estimates predicted a worsening world energy crisis, with the Soviet

Union, acting aggressively to protect its energy supplies. The Carter administration bestowed power on a new generation of energy philosophers and decision makers, espousing energy conservation and renewable energy use at the expense of conventional energy supplies, especially nuclear energy. To demonstrate his commitment to new forms of energy, President Carter ordered that the reviewing stand for the inaugural parade be heated by a solar hot water system. Ten thousand square feet of collectors, supplied by ERDA, were installed temporarily on the White House lawn behind the reviewing stand to supply the solar-heated water that would then be pumped to radiators in the stands. Although the system, for a variety of reasons, performed poorly, the mere fact that it had been requested bode well for solar energy. The energy wheel had turned 180 degrees.

President Carter made it known almost immediately that James R. Schlesinger, with a broad background in senior government positions, would be his chief energy strategist. Schlesinger quickly assembled several task forces to develop the administration's energy policies. Following up on his election rhetoric, and repeating President Ford's recommendation made just two months earlier, President Carter, on March 1, 1977, proposed to Congress the creation of the cabinet-level Department of Energy, which would oversee almost all of the government's energy activities, from RD&D to regulation.

The early energy activities of the new administration focused on developing a National Energy Plan (NEP) and preparing the president to address the nation on the nation's energy problems and their solutions. It fell to Alvin L. Alm, former EPA Assistant Administrator for Planning and Management, selected by Schlesinger as one of his principal assistants, to oversee development of the NEP. The detailed outline of a national energy plan was put together in just under two months. Goals to be achieved by 1985 focused on the reduction of energy demand and oil imports. Among the three strategies outlined to achieve the goals was "a vigorous research and development program to provide renewable and essentially inexhaustible resources to meet U.S. energy needs in the next century."

Energy conservation, in all sectors of the economy, was to be the cornerstone of the plan. Solar energy use in the buildings sector was to be an important element, with up to $100 million in spending promised in the next three years to add solar hot water and space heating to suitable federal buildings. Various types of tax credits were also proposed to stimulate the development of a large solar market.

On April 20, 1977, three months after his inauguration, President Carter presented his NEP outline to a joint session of Congress and a large national television audience. He announced a national goal of 2.5 million homes to be equipped with active solar heating and cooling systems and tax credits for home owners installing such systems. With solar energy contributions identified in many sections of the outline, the ERDA solar staff saw growth in all programs.

In May 1977 ERDA proposed the idea of putting a solar heating and hot water system on the White House, and approval was given for a feasibility study. What greater endorsement could there be than to advertise that the First Family was bathing in water warmed by the sun? By the end of July 1977, the feasibility studies had been completed, and an internal debate began on whether such a system should be installed. This debate carried over into 1978, when Schlesinger intervened by appointing a task force to resolve issues such as what type system should be installed and where, and which agency—DOE, GSA, or the National Park Service—should be in charge of the project. Based on the task force's recommendations, three alternatives were submitted to Hugh A Carter, Jr. special assistant to the president for administration, at the White House at the end of February. It would be another seven months before final approval to proceed was given.

While the Carter NEP was attracting the national spotlight, vigorous politicking was underway among the agencies to be incorporated into DOE to protect programs and positions. At ERDA, which would bring the bulk of the program funds and personnel, to DOE, senior management struggled with many unknowns. In February 1977, Robert Seamans resigned and Robert Fri was designated acting administrator. Robert Hirsch, on the job for only one year after replacing John Teem, also announced his departure. Other ERDA senior managers, appointed by a Republican administration, also began to leave; those who remained carried on without much clout or influence on the Schlesinger team. Many, including Fri, made it known they would be willing to stay if asked, but as the spring of 1977 advanced, it became increasingly unlikely that this would happen.

PL 95-91, establishing DOE, was signed into law on August 4, 1977. As expected, Jim Schlesinger was named Secretary of Energy, and he brought with him most of the senior staff and advisors he had assembled earlier in the year. The final language of PL 95-91 created major changes in program responsibilities: the president was to appoint five assistant secretaries and one director, with titles and responsbilities often completely different from those of the ERDA assistant administrators. This necessitated a painful shuffle of programs and personnel from the ERDA assistant administrators offices to new DOE assistant secretaries.

Nowhere were these changes more painful than in solar energy programs. RD&D responsibilities were to be split between two assistant secretaries. Whereas the ERDA Assistant Administrator for Solar, Geothermal and Advanced Energy Systems had been responsible for the evolution of all solar technologies from research through commercialization (FEA's mandated responsibilities notwithstanding), Solar Heating and Cooling of Buildings and Agricultural and Industrial Process Heating were split off from Solar Electric Technologies. Responsibility for RD&D, commercialization, and application for Solar Electric Technologies was given to the Assistant Secretary for Energy Technology. Corresponding responsibilities for Solar Heating Technologies were given to the Assistant Secretary for Conservation and Solar Applications. Of particular note, both these offices would share in the responsibility of funding and oversight of the newly created SERI. Potential overlap also existed with the Assistant Secretary for Resource Applications, whose responsibilities included energy supply commercialization, which sounded suspiciously similar to some of the previously mentioned activities. Another major change was the reporting chain. At ERDA, all assistant administrators reported directly to the administrator. With many functions now agglomerated within the new department, the major R&D offices or "Outlay Programs" were to report directly to the under secretary, who on the management pecking order held the number three position, behind the secretary and deputy secretary. The deputy secretary was given the major responsibility of overseeing the regulatory and energy information functions.

In preparation for DOE's inauguration on October 1, 1977, ERDA's staff started identifying people and programs to be allocated to the new organizational entities. But amid the confusion and uncertainty there was also some good news. Dale E. Myers, president of Jacobs Engineering and former NASA Assistant Administrator for Manned Space Flight, was nominated to be the first under secretary. Myers was a manager well known to some of the senior ERDA staff, and he seemed admirably suited to these new responsibilities. All solar technologies would report to him through their two assistant secretaries.

On September 30, 1977, Schlesinger chaired the first meeting of his senior staff, which he named the Energy Policy Council. Present were Deputy Secretary designate John O'Leary, Dale Myers, and other senior staff, most still in an acting or unconfirmed status. Schlesinger described his vision for DOE: to succeed, the department must be "outward looking" and "engage the enthusiasm of the public"; it must act, not as the individual constituents of its former parts, but as a single entity. O'Leary agreed that this would be the hardest job for the new DOE management. The Carter White House, with Schlesinger as its voice, had made energy the primary concern of the new administration. In the weeks and months ahead, Schlesinger was the most visible member of the Carter cabinet carrying the energy message to Congress, the media, and the public. A key part of his message was that the energy problems facing the nation would intensify in the 1980s, and that the nation would have to change its energy habits. Robert D. Thorne, formerly head of the AEC's San Francisco office and Acting Assistant Administrator of the Nuclear Energy office during ERDA's last months, was nominated and eventually confirmed as Assistant Secretary for Energy Technology. Omi Walden, who had been the head of the Georgia State Office of Energy Resources for then Governor Carter, was nominated as Assistant Secretary for Conservation and Solar Applications, the second part of the solar energy management team. Her nomination, in November 1977, became bogged down in partisan wrangling and the author, then acting assistant administrator in ERDA's last months, became acting assistant secretary while awaiting Walden's confirmation.

The early months of DOE were a time of consolidation and "getting acquainted." In addition to those from ERDA, programs and personnel from FEA, DOC, DOD, ICC, DOI, and HUD were now part of the new department; field organizations, national laboratories, and power marketing administrations were also folded in. Directors of what, in many cases, had amounted to private fiefdoms scrambled to understand who their new bosses would be and what new ground rules they would need to observe. Real progress in the "Outlay Programs" slowed while all the changes were being digested. In the new office of the Assistant Secretary for Conservation and Solar Applications, the solar energy programs were consolidated under the Assistant Director for Heating and Cooling; Ronald D. Scott, brought into DOE from ERDA, became the first Director. Although the president's National Energy Plan had not been enacted in 1977, the direction of the compromises that would be needed for the plan to be accepted became clearer, and it was becoming more evident that the new administration would strongly support increasing the scope of the solar programs.

During DOE's first months, the staff began a difficult period of adjustment to a new way of doing business. The FY 1978 budget, to a large degree an amalgamation of budgets from many different, formerly independent agencies, was in disarray. Supplemental budget requests were being formulated and sent to congressional committees, and a number of project deferrals were being proposed by OMB. The FY 1979 budget cycle was in full swing, complicated by a new administration tactic called "zero-based budgeting" (ZBB), which forced all programs to restructure and redefine their budget line items in new terms. No one really knew how to respond to this OMB guidance, and long delays became common in reviewing and approving budgets with Under Secretary Myers, in preparation for transmittal to OMB. The solar programs were not immune to these problems, and the reorganization of solar programs under two assistant secretaries only added to the burden. The bright side of the story was that the Solar Heat Technologies budgets continued to expand. The FY 1979 request to OMB grew from the congressional appropriation of approximately $206 million in FY 1978 to $220 million, with supporting arguments that would add an additional $75 million to fully fund the goals of the NEP.

With his background at NASA, Dale Myers was eager to continue receiving NASA's support for energy programs. The MOU signed by ERDA and NASA in June 1975 was revisited and modified to reflect DOE's responsibilities. The new MOU, signed by Myers and NASA Deputy Administrator Alan M. Lovelace on March 21, 1978, established the DOE-NASA Program Coordination Committee, which began to meet quarterly; under its terms, all programs performed by NASA for DOE were to be budgeted by DOE, with the funds transferred to NASA through interagency agreements (task orders). With the new MOU in place and clear lines of authority established, both DOE assistant secretaries responsible for solar RD&D programs began to transfer the management of many contractor activities to NASA centers. For Solar Heat

Technologies, the Marshall Space Flight Center became the major support center.

A new complication was added to the management of the solar programs with the unexpected appointment of Jackson S. Gouraud as Deputy Under Secretary for Commercialization. His office came with an ill-defined charter that included all energy technologies and thus overlapped other offices' responsibilities. Although he reported directly to Under Secretary Myers, Gouraud had no staff with which to implement his proposals, forcing him to borrow staff from already overburdened programs. The new deputy under secretary quickly foresaw the improving climate for solar technologies and the opportunity to make an impression on the Carter administration; he became another voice calling for the rapid commercialization of passive solar designs and solar hot water systems, based on large-scale, federally sponsored demonstrations and other federal incentives.

With the earlier congressional mandates to commercialize solar energy as background and the new focus coming from Under Secretary Meyers's office, a working group was established in the spring of 1978 to develop a commercialization plan for solar water heating. (Passive solar commercialization was deferred for the time.) By October 1978, when the working group released its "Commercialization Plan for Solar Water Heating," HUD had issued almost 11,000 homeowner grants of $400 each for single-family hot water installations in the Northeast states and Florida. The widespread response to the HUD program appeared to justify optimism that the public at large was now ready to accept and even embrace solar hot water systems. The cost of solar hot water systems for residential use had been dropping. In 1978 pumped systems cost from $1,200 to $3,000, and thermosiphon systems from $500 to $1,500, depending on user needs; both of these costs included installation. For a small thermosiphon system, the HUD grant had the potential of defraying most of the initial system cost.

The commercialization plan had five major elements: (1) market stimulation, (2) information dissemination, (3) industry support, (4) technology improvement, and (5) overall program support (a catchall that included funding to SERI and the RSECs, market analysis, and program evaluation). The estimated cost of the program from FY 1979 through FY 1985 was $65.4 million. The plan envisioned that by 1985, 2 percent of the total potential market would be captured, leading to a .06-quad energy savings, and that by the year 2000, a 0.5-quad savings would be realized. The 1985 projected savings was identical to that projected in 1977 by the MOPPS study and far below all other projections.

Reports supporting the increased use of solar energy continued to proliferate. In April 1978, the President's Council on Environmental Quality (CEQ) published *Solar Energy: Progress and Promise.* Assembled with little assistance from the DOE program offices, this report concluded that solar energy could contribute up to 25 percent of the nation's energy needs by the year 2000. To achieve this very large contribution, the CEQ proposed a proactive government program that included large federal purchases, as well as financial and tax incentives, and mandated price increases for conventional energy sources. In May, the Office of Technology Assessment (OTA) published *Application of Solar Technology to Today's Energy Needs.* Critical of the ERDA/DOE emphasis on large-scale solar energy systems, the OTA report recommended that solar RD&D programs concentrate instead on small-scale, direct-energy systems and that the federal government promote the use of these systems by reducing their price—through incentives to purchasers, large-scale federal purchases, and manufacturer incentives. OTA also recognized the need to develop performance standards to address the growing problem of shoddy manufacturers and installers. In June, DOE published a slim volume *Solar Energy: A Status Report*, which took at face value the CEQ and OTA projections, as well as those of the many other pro-solar reports, lending its voice to the chorus of solar energy supporters who saw in the Carter administration an opportunity to advance a new energy agenda for the nation.

On May 3, 1978, in his "Sun Day" speech at SERI in Golden, Colorado, President Carter called for a Domestic Policy Review (DPR) of solar energy. He cited the CEQ projection that a quarter of the nation's energy needs could be supplied by solar energy; the DPR was to develop an overall strategy to meet that projection.

The DPR of solar energy became a cabinet-level review chaired by Jim Schlesinger. On May 16, in a memorandum addressed to all the cabinet secretaries as well as the administrators and directors of key independent agencies, Stuart Eizenstat, writing for the president, laid out the purpose and structure of the review. Recommendations were to be submitted to the president by September 1, 1978. By means of the DPR, solar energy had been elevated to the highest councils and awareness of government.

Jointly led by White House Domestic Counsel Kathryn P. Schirmer and Al Alm at DOE, the DPR of solar energy involved many different agencies in the executive branch and heavily committed the resources of three: DOE, HUD, and NASA. Public meetings were held in a number of cities across the country where DOE officials as well as local and state officials, industry, and ordinary citizens discussed and debated how to accelerate the acceptance of solar energy systems into the nation's energy economy. The agencies involved set up large task forces to develop the backup data and to sort through the information from the public meetings. The sheer size of the DPR effort eventually forced a postponement of the due date established by Eizenstat.

In August, Schirmer rejected the second status report of the DPR findings. DOE's solar offices, although working diligently through the summer, felt they had been deliberately excluded from the inner circles that created the report's conclusions. The timing of the public statement and the separate response memorandum to the White House seemed to confirm fears that public and solar advocate positions, as well as solar expert advice, had been ignored or kept to a minimum. Indeed, when the draft DPR was released for public review in September 1978, it received harsh criticism from solar activists. Though comments had been solicited from the public, the replies were not received in time to be factored into the final response memorandum that went to the cabinet department secretaries and then to the president. Because the report was already late in getting to the president, public comments could only be appended to the main body of the DPR.

The end result was another highly optimistic projection of the potential impact of solar energy on the nation's energy consumption. The "Domestic Policy Review of Solar Energy, a Response Memorandum to the President of the United States" was released in February 1979. Nine major findings were listed. The primary finding was that, considering the contribution already made by biomass and hydropower (4.8 quads in 1978), solar energy could displace 10–12 quads of a projected total of 95–114 quads by the year 2000. This would be achieved from a base case scenario defined by existing policies and programs and with certain assumptions on the price of energy. Even larger impacts, up to 30 quads, might be possible with massive federal intervention. To achieve these ambitious goals, three options were outlined for the president:

• Option 1 was to continue current programs but make them more effective. It did not require new legislation, but asked for $50 million of the FY 1979 budget for solar RD&D to be redirected and for $225 million in additional funds (that is, above those currently projected) in the years 1980-1985. Option 1 supported a solar energy budget of $742.3 million for FY 1980, an increase of $250 million over FY 1979. This option stressed solar heating, with its projected nearer-term impact, and de-emphasized solar electric applications.

• Option 2 added to the elements of option 1 proposals for a new Solar Energy Development Bank, a 30 percent tax credit for industrial and agricultural applications, a set-aside by the Rural Electrification Administration of more of its loans for solar energy systems, more solar equipment on federal facilities, and tax credits for passive designs and for equipment that had been certified by independent testing laboratories under government supervision. The cost of option 2 was estimated to be $2.4–2.6 billion over that of option 1 between 1980 and 1985.

• Option 3 added to the elements of options 1 and 2 proposals that would dramatically increase federal support for solar energy. It required 10 percent of new electrical generating capacity to be from renewable resources by 1985, a 60 percent tax credit for builders of energy-efficient houses, a 40 percent tax credit for industrial applications and biomass (twice the level of the new energy tax law in effect at that time), and establishment of a Solar Energy Development Bank with interest-free loans for residential and commercial customers and 3 percent loans for manufacturers. The cost of option 3 was estimated to be $5.6 billion over that of option 2 between 1980 to 1985.

Because of the haste in which the final report was assembled, it contained a number of confusing statements and internal inconsistencies. At hearings called by the House Commerce Energy and Power Subcommittee, a large number of witnesses testified. Among them was George Löf, one of the original NSF grantees and ERDA contractor and now chairman of the board of Solaron Corporation and senior advisor of Colorado State University's Solar Energy Laboratory. Löf objected strongly to DOE's solar program priorities. He stated that money applied to the development of methods and hardware for solar electricity generation, as opposed to the production of solar heating systems for building and manufacturing, ignored the nation's needs. He stated that the Solar Heating and Cooling Demonstration Program had been conceived with unrealistic objectives and unfortunate consequences. Instead of demonstrating proven solar heating systems, Löf said, it had primarily been a testing program for untried, speculative, and expensive ideas proposed to the government by inexperienced designers and manufacturers. The program consequently misled the public into thinking that solar heating might not be a successful application after all. Löf stated that a different public opinion could have resulted if the solar heating program had been properly designed and implemented.

California Energy Commissioner Ronald Doctor testified that the DPR failed to promote the maximum economic use of solar energy in the near and the long term. He stated that the DPR was deficient in not tapping the utility sector for lending capital, mandating solar uses where cost-effective, developing a solar strategy for urban tenants, or ensuring adequate consumer protection through a competitive and responsive solar industry.

Richard Munson, coordinator of the Solar Lobby, testified that the DPR was long on rhetoric and short on substance. For example, there were plans for a Solar Energy Development Bank, but funds would not to be available to run it until FY 1981. Munson claimed that the DPR only called for a slight increase in the base funding of solar work for the following fiscal year—well below the amount requested.

Solar advocates skeptical of the DPR began to draw up a "Blueprint for Solar America," which recommended option 2-1/2 (a program that would fall between options 2 and 3) and, in addition, required blending gasohol into gasoline, consideration of passive solar in all new building designs, and federal tax incentives in addition to any state tax benefits that might be available for the installation of solar systems.

DOE, after little more than a year of operation, could look to at best mixed reviews on its overall accomplishments. Consolidating regulatory and RD&D functions, not to mention weapons programs, under one roof had turned out to be a major undertaking, with the potential to become unmanageable. As predicted by some, decisions on politically sensitive regulatory functions were monopolizing the attention of senior management, to the detriment of the energy research programs. Egged on by an increasingly vocal and strident solar and environmental lobby, and bolstered by strong support from the administration and from powerful congressional leaders and their staffs, the department found itself pushing solar energy programs to promise more than they could deliver.

Notes

1. Ford Foundation, Energy Policy Project, *A Time to Choose: America's Energy Future* (Cambridge, MA: Ballinger Publishing Co., 1974).

2. H. Guyford Stever, Director, National Science Foundation, letter to Olin E. Teague, Chairman Committee on Science and Technology, U.S. House of Representatives, Washington, DC, April 4, 1975.

3. *ERDA-48*, "A National Plan for Energy Research, Development and Demonstration: Creating Energy Choices for the Future," vol. 1, June 28, 1975.

4. *ERDA-49*, "National Solar Energy Research, Development and Demonstration Program, Definition Report," June 1975.

5. Olin E. Teague, Chairman, Committee on Science and Technology, U.S. House of Representatives, letter to H. Guyford Stever, Director, National Science Foundation, March, 5, 1975.

6. *ERDA-23*, "National Plan for Solar Heating and Cooling (Residential and Commercial Applications): Interim Report," March 1975.

7. *ERDA-23A*, "National Program for Solar Heating and Cooling," October 1975.

8. ERDA, *An Executive Analysis of Solar Water and Space Heating*, DSE-2322-1, November 1976.

9. National Academy of Sciences, National Research Council, Solar Energy Research Institute Committee Report, *Establishment of a Solar Energy Research Institute*, September 1975.

10. MITRE Corporation, *Industry Opinions on the Formation of a Solar Energy Research Institute*, Technical report MTR-7067, October 1975.

11. *ERDA-76-1*, "A National Plan for Energy Research, Development and Demonstration: Creating Energy Choices for the Future," 1976.

12. Roger H. Bezdek, personal communication.

13. U.S. Department of Energy, *Market-Oriented Program Planning Study (MOPPS): Final Report*, December 1977.

14. An interesting analysis of the political and policy consequences of the MOPPS can be found in Aaron B. Wildavsky, Ellen Tenenbaum, and Pat Albin, *The Politics of Mistrust: Estimating American Oil and Gas Resources* (Beverly Hills, CA: Sage, 1981), which includes some detailed analysis of the MOPPS and its impact.

5 The Growth Years: 1977–1980

Frederick H. Morse

At this point, our account of the history of the solar heat technologies program changes authors and perspective. Chapters 5 and 6 focus on the people, events, legislation, and decisions that had specific impacts on the solar heat technologies program, as opposed to the broader solar energy program. It is also now almost exclusively a history of the Department of Energy's Office of Conservation and Solar Applications and the Office of Solar Heat Technologies, which the author directed until July 1989, and from which the work described in the following volumes of this series was managed and supported.

The main themes of chapters 5 and 6 are administration policies and how they were implemented by the various Assistant Secretaries for Conservation and Solar Applications, the budget battles that raged between the administration and Congress, the organizational changes that occurred within the department and their impacts on the technical programs, and the development and evolution of the field support structure of laboratories, administrative offices, and universities. These themes set the context for the major RD&D and commercialization activities of the solar heat technologies program, and for its interactions with the solar energy industry and the various solar advocacy groups.

5.1 The Carter Administration Settles In

Although President Carter moved quickly after his inauguration in January 1977 to establish the Department of Energy and appoint its top officials, the effect of these changes on the solar heat technologies program was not to be felt for many months. Delay in congressional confirmation of Carter appointees meant that for the first twenty months of the Carter administration, the program's policies, planning, budgets, and, perhaps most important, management, established under NSF and ERDA, continued without change.

From DOE's inception in August 1977 until August of 1978, the Acting Assistant Secretary for Conservation and Solar Applications was Donald A. Beattie, who had been solar energy program administrator for NSF and ERDA. Under Beattie's guidance, the program had developed strong RD&D plans for buildings and industrial applications, set up a field support structure, and initiated RD&D programs through government laboratories, field offices, and numerous contracts with and grants to industry and universities. Funding for the program was in place and growing rapidly, and most of the managers at the new DOE knew their RD&D goals and seemed to have control of the technical and financial resources necessary to accomplish them. But the program was about to undergo major changes.

In the late summer of 1978, with more than a year under its belt, DOE's solar heat technologies program was poised to capitalize on the rapidly improving climate for renewable energy programs. The Carter administration was solidly behind increasing the budgets for these programs, which were also supported by Congress and public opinion. The major DOE organizational issues had generally been resolved and the mechanics of making a new, large, cabinet-level organization function responsively to both internal and external constituencies were being better understood with each passing day. Problems still remained, but essentially all of the senior staff was in place and lines of communication established. For solar heat technologies, the final key management position, the Assistant Secretary for Conservation and Solar Applications, was about to be filled.

5.1.1 The Appointment of Omi Walden

After a long, at times difficult, confirmation process, Omi Walden finally was confirmed in August 1978 as Assistant Secretary for Conservation and Solar Applications. During her confirmation hearings, Walden defined the framework for her office at DOE, stating that solar technologies must be aggressively transferred into the marketplace at prices that were competitive with traditional energy systems. With this, she signaled her intention to emphasize DOE's role in wide-ranging commercialization activities. Throughout the Carter presidency, the issue of government involvement in commercialization was the focal point for a great deal of turmoil in the solar-related programs of the Department of Energy.

In addition to her job as head of the Georgia State Office of Energy Resources, Walden had been Carter's advisor on energy and environmental policy during his tenure as governor. Walden's skills and preferences were in developing policy rather than in the technical aspects of solar energy and energy efficiency. A master at politics and negotiation, Walden worked well with the OMB and members of Congress; she had a facility for discerning potential long-term benefits of solar technologies and for understanding public opinion. Walden's strengths were in realizing how critical it was to get solar technologies into the marketplace and in conveying her vision to staff. She also recognized that many technical problems had to be solved and that all of the solar technologies needed further research and development. During Walden's tenure as assistant secretary, the solar energy and conservation program budgets continued to expand.

Walden worked to change the focus of the DOE program from R&D to commercialization. Determined to strengthen the association of solar technology with conservation issues, Walden directed her staff to emphasize passive solar heating and cooling applications. She also believed DOE's programs should promote whole building systems that incorporated both conservation and solar measures, rather than either one exclusively. For the first time, DOE combined these areas in one office.

Coming to Washington, D.C., from a small office in Georgia, where she managed a staff of twelve, Walden was suddenly in charge of an organization with a staff of over 300 and a $1 billion budget. It soon became evident that her desire to be involved in the details of the activities of her office was not very productive in managing such a large organization. In addition, Walden faced an upper management that was pro-nuclear and not sympathetic to solar technologies. Walden's first efforts concentrated on organizing and staffing her office, actions that had been held in abeyance for months pending her confirmation.

5.1.2 Reorganization of the Office of Conservation and Solar Applications

When Walden took over the solar and conservation programs, she found a staff with strong technical backgrounds consisting primarily of people who came to DOE from ERDA, many of whom had earlier backgrounds at AEC, NSF, NASA and FEA. Most of the managers on her staff were experienced with federal R&D programs but had little experience with market development or contact with the states. When completing the staffing of her own office, Walden favored young people from outside the federal government who believed in the future of solar applications and upon whom she could rely to support her aggressive commercialization plans.

Her first organizational change was to create three new high-level positions: an executive director and two deputy assistant secretaries. Director of the Office of Buildings and Community Systems Maxine Savitz was selected to be the Deputy Assistant Secretary for Conservation, and Kelly Sandy from Walden's staff became the Acting Deputy Assistant Secretary for Renewable Energy and Executive Director of the Office of Conservation and Solar Applications.

In January 1979 the House Committee on Oversight and Investigations released a report concluding that DOE's solar demonstration programs were not as effective as they could be. According to this report, many proven solar technologies were not being demonstrated because the federal government had not encouraged the best use of solar technologies and had not inspired consumer confidence in these technologies. The report claimed that the focus of solar demonstrations had been on the installation of the more expensive and complex active solar heating and cooling systems; furthermore, that passive solar designs had not been promoted, tax credits had not been offered for passive solar installations, and information on the performance of the demonstrated solar technologies had not been made available to the general public. Omi Walden's reorganization of her office and redirection of the solar program was designed, in part, to address the problems identified in the House report and in the soon-to-be released Domestic Policy Review of solar energy, even though many of the findings of these reports were open to debate.

Walden further responded to the House report and the DPR by increasing the funding for DOE's commercialization programs and by developing the National Plan for Accelerated Commercialization of Solar Energy. She formed a DOE working group to determine the market readiness of specific solar technologies and what was required to achieve full market readiness. Walden said that achieving the goal of equipping 2.5 million homes with solar energy systems, proposed by President Carter in his energy address to Congress (April 20, 1977) was contingent on passage of the National Energy Act, supporting actions by DOE, and tax credits. Today, even though solar tax incentives were phased out and DOE's funding for active and passive solar energy systems decreased dramatically in the 1980s, it is estimated that over 1 million homes either are equipped with solar water heaters or incorporate passive solar designs.

Deputy Under Secretary for Commercialization Jackson Gouraud had conducted a review of many of DOE's R&D activities and selected four solar technologies that DOE would try to commercialize immediately. These were solar water heaters, passive solar heating, photovoltaics, and parabolic troughs, all of which became part of Walden's program during 1979. As Walden was responsible for moving these solar technologies into the marketplace, she established the Office of Commercialization, which carried out the work assigned by Gouraud. The Office of Solar Applications was responsible for the active and passive solar, agricultural and industrial process heat (which included parabolic trough solar collectors), and photovoltaic systems R&D programs, as well as for funding the defined commercialization activities related to those technologies.

5.1.3 Commercialization and Market Development

Prior to Omi Walden's arrival, the solar energy program, developed under ERDA, focused on RD&D. Under Walden, the program became more directly involved in commercialization and linked more closely with energy conservation. Walden believed that DOE's solar commercialization programs would fail unless the states were significantly involved in their definition and implementation. She stressed grassroots involvement in promoting solar heating and cooling. Having come from a position reporting to the governor of Georgia, she had a network of important contacts from the National Governors' Association Energy Committee, through which she intended to promote decentralized management of solar programs.

Getting solar energy systems to market became the biggest challenge faced by DOE. Most technologies were still at an RD&D phase. The mission of the Office of Solar Applications was to bring about rapid changes in the marketplace that would result in people buying systems that used solar energy rather than conventional energy sources. DOE's program now required new elements to provide support for training solar energy system installers and maintenance workers, for expanded information and education programs, and for developing codes and standards to assure consumer satisfaction. During her time in office, Walden promoted the prospect of jobs creation through the greater use of solar energy. One way she hoped to accomplish this was by establishing special solar centers in the states for training people in the installation and repair of solar technologies. In addition to these new program elements, DOE continued to fund the R&D essential for the further evolution of the various active and passive solar technologies.

The thousands of solar demonstrations required by Congress and supported by the government prior to Walden's arrival at DOE had shown mixed results. Furthermore, many people believed that government-supported demonstrations distorted the market by subsidizing those technologies. In a major policy change, Walden's staff began the phaseout of the demonstration programs that had characterized the solar heating and cooling effort mandated by Congress. No new demonstration cycles would be initiated and those underway would be brought to a conclusion. The challenge was to determine an effective way to help develop the market for solar products without distorting the marketplace. In place of demonstrations, the Office of Solar Applications created a program strategy for federal support that paralleled the decision-making path taken by equipment manufacturers in the private sector (see figure 5.1). In this program, DOE would share the cost of testing solar energy systems in prime markets identified by participating companies. DOE would concentrate its efforts and funding on the first four steps of the development process—i.e., basic research, market research component development and testing, and prototype systems development and testing—and subsequently on product improvement and the removal of market barriers.

According to this strategy, DOE cost-shared funding would help industry move new ideas and concepts from the research phase into the development phase, where laboratory tests would be performed on a few units. Engineering field tests of typically 1 to 10 units would follow to establish the technical and economic effectiveness of solar products. Then, with the companies assuming the greater share of the costs, additional units, typically 10 to 100, would be test-marketed in prime applications, with the intention of establishing performance as well as maintenance and service records. After that, DOE would provide minimal assistance for the production of proven and reliable products, supported by appropriate private sector marketing activities.

During the period of the large demonstration programs, ERDA had projected that for every solar energy system installed with government funding, the private sector would sell and install at least ten systems. However, during the demonstration program, ERDA, then DOE, installed far more systems than the private sector did. (At the time of the publication of this volume, this trend had changed and the private sector more than exceeded the 10:1 ratio in the residential and commercial applications of solar water heating.)

The Growth Years: 1977–1980

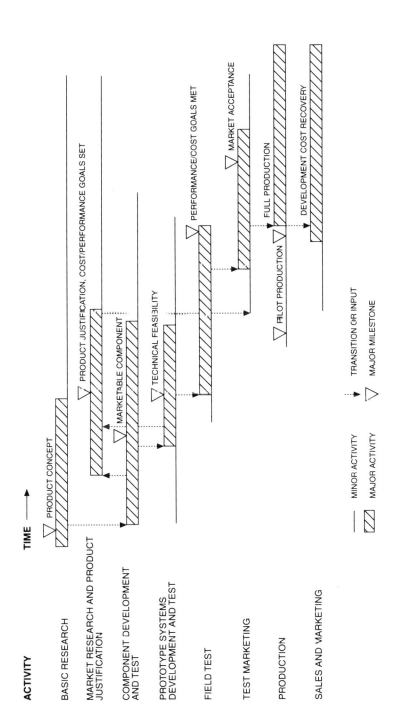

Figure 5.1
Government functions in support of the product development process.

While there were no plans for additional demonstration cycles, DOE did plan to support some market testing. Some of this was to be accomplished through the Solar Federal Buildings Program, which had been restructured to support the field-testing of advanced solar products and systems, rather than as a demonstration program to show the readiness of these technologies.

5.1.4 The Solar Federal Buildings Program

In April 1976 the Federal Energy Administration released guidelines and an implementation plan for its proposed Solar Federal Buildings Program (SFBP). The objective of this five-year project was to accelerate the commercialization of solar heating and cooling in the United States. The program would use federal buildings to provide a significant, early market for solar hot water and space heating equipment, which then was to be incorporated into the planning and designs of all new federal buildings and major renovations of older federal buildings.

However, when this program moved into DOE, Secretary Schlesinger was reluctant to endorse the installation of solar energy equipment on federal buildings, believing that these technologies were not mature enough. Schlesinger argued that Carter favored DOE encouraging the voluntary acceptance of these technologies and that solar installations should be gradually mandated on federal buildings as the technology developed. When DOE published the proposed rules for the SFBP early in 1979, inviting written comments, the American Institute of Architects (AIA) criticized them, expressing concern that the technology still needed further investigation and that expensive or poorly operating solar technologies would be installed. Moreover, AIA thought that DOE, not NASA, should oversee the program. (While NASA initiated the SFBP under direction of DOE, the responsibility was moved to the DOE San Francisco Operations Office when NASA terminated its solar energy activities in 1984.) SEIA took issue with the proposed requirement that each project have a one-year warranty from the installer and a five-year warranty from the manufacturer, claiming that such requirements were too stringent for a new and emerging industry such as solar. The final rules were published in October of 1979.

Although by January 1979, DOE already had a well-developed implementation plan and the solar industry was ready to participate, everything was put on hold by OMB, which did not like the program's high cost (around $100 million). In October of 1979 the official Request for Proposals (RFP) was issued. From the 912 proposals evaluated, in May 1980 DOE selected 843 for awards totalling $31 million to sixteen different federal agencies, with projects in all fifty states and the District of Columbia. During the course of the program, 142 of these projects were withdrawn or canceled for technical, cost, or schedule reasons, for contractor default, or at the request of the proposing agency; because some projects included more than one system, 718 systems were built over the course of the program.

The technical results of the SFBP are reported in some detail in volume 10 of this series. The authors conclude that the program achieved limited market stimulation through the installation of over 700 systems, federal agencies did start to employ solar technologies (and more would have had the incentives continued), design, installation, operation, and maintenance experience objectives were met, and good applications of solar energy were identified. As a result of the lessons learned during the program, the American Society of Heating, Refrigerating, and Air-Conditioning Engineers (ASHRAE) published a set of comprehensive design, installation, and maintenance manuals for active solar heating and hot water systems.

Other program objectives, however, were either not met or marginally met as a result of the changing political atmosphere, the falling oil prices, and reduced budgets and lack of motivation on the part of some agencies. Consumer confidence in solar energy increased only minimally; economies of scale as a result of large procurement were never realized. There was no major shift to solar energy, and industrial/federal agency interaction was stimulated only to a small degree.

5.2 The Carter Administration's Recommitment to Solar Energy

5.2.1 The Dedication Speech

At the June 20, 1979, dedication of the White House solar hot water heater,[1] more than two years after he launched the Domestic Policy Review (DPR) of solar energy and a half year after he received the DPR Response Memorandum, President Carter announced a goal of 20 percent of the nation's energy needs to be met in the year 2000 by solar energy resources. Warning that his 20 percent solar goal would not be easy to accomplish, the president stressed the need for a variety of initiatives, including the creation of a Solar Energy Development Bank, as well as additional tax credits for builders of passive solar homes, installers of solar industrial process heat systems, and buyers of airtight wood-burning stoves. The president also issued a number of directives to federal agencies to increase their use of solar technologies, and to initiate more pilot projects that would demonstrate the effectiveness of those technologies. Carter also recognized the importance of educating the American public by demonstrating that solar technologies work.

One of the most controversial aspects of the White House solar dedication speech was Carter's support for a Solar Energy Development Bank, (hereafter referred to as the "Solar Bank"). Proposed at an initial level of $100 million in FY 1981 and at $400 million for the following three years, the Solar Bank's low-interest, long-term loans would be financed by money from a proposed windfall profit tax on oil company earnings and from a proposed energy security trust fund (which never materialized). The Solar Bank would provide up to $10,000 financing for a solar energy system in a single-family residence, $5,000 for each unit in a multifamily dwelling, not to exceed a total of $500,000 per dwelling, and $200,000 for a solar energy system in a commercial structure.

Reaction to the president's speech was mixed. Many pro-solar organizations and supporters were pleased with Carter's commitment, but some did not think he went far enough in setting a national strategy for the promotion of solar energy. The Environmental Policy Center said that if solar was so good, it deserved endorsement beyond the Solar Bank's trust fund. And SEIA President Sheldon Butt called the Solar Bank's financing approach a bad idea.

5.2.2 The Purge of the Carter Cabinet

In July 1979, with his popularity and political prospects on the wane, President Carter replaced many of his cabinet members. DOE Secretary James Schlesinger was replaced with then Deputy Defense Secretary Charles Duncan. Carter's removal of the outspoken, first DOE secretary was viewed by many to show that energy policy would be determined more than ever by the White House; Carter seemed to be looking for someone to implement his own policies more assiduously.

In spite of her strong advocacy of renewable energy, Assistant Secretary Walden ran into problems when former Federal Energy Administrator and energy czar John Sawhill was named to replace Deputy Secretary John O'Leary in October of 1979. Walden, who earlier had lost a few turf battles and other bureaucratic skirmishes in DOE, was seen as a scapegoat for some unsuccessful programs undertaken by DOE (described below). The new management team of Duncan and Sawhill began to consider a number of personnel changes, resolving to replace Walden with her colleague from Georgia, Thomas Stelson, Dean of Engineering at Georgia Institute of Technology. Because of Walden's close ties to President Carter and OMB budget chief McIntyre, such a change was not easy to implement. Stelson was reluctant to take the position until another position could be found for Walden; Walden was therefore placed in a newly created position of Advisor to the Secretary for Conservation and Solar Marketing. As Walden's top priority had always been these marketing efforts, DOE hoped that she would elevate the visibility and success of the department's solar market development efforts.

In October 1979, shortly after Stelson's appointment, DOE decided to recombine all renewable energy programs into the Office of the Assistant Secretary for Conservation and Renewables. All of the solar and renewable R&D functions of the Office of Energy Technology were combined with the solar heat technologies R&D and commercialization efforts of the Office of Conservation and Solar Applications.

As was the fashion for all new assistant secretaries, Stelson began by reorganizing his office. Maxine Savitz, who briefly served as acting assistant secretary prior to Stelson's appointment, continued as Deputy Assistant Secretary for Conservation. The Solar Applications Office was divided into three new offices: Solar Applications for Buildings, Solar Power Applications, and Solar Applications for Industry. Solar Applications for Buildings included the active, passive, and photovoltaic technologies, and was headed by the author of this chapter. Placing photovoltaic (PV) technologies into a buildings organization was interpreted as an indication that DOE intended to place a higher priority on distributed PV installations, rather than on central station generation. Solar Power Applications included wind energy conversion systems, solar thermal energy conversion and ocean thermal energy conversion. Solar Applications for Industry included industrial process heating and biomass conversion.

Robert San Martin was appointed as Deputy Assistant Secretary for Solar Energy, with the three renewable energy offices reporting to him.

San Martin had a strong academic background in solar energy R&D, and thus became the most senior DOE official with a background in solar technology. San Martin assumed his position at a time when the overall federal involvement in solar energy spanned many agencies in addition to DOE, and included funding of about $1 billion per year.

San Martin identified four mechanisms for governmental action to advance solar energy: information dissemination, regulation, financial assistance such as loans and grants, and programmatic research and development (the original base of solar work in the early 1970s). He felt that the challenge was to determine which mechanisms to apply and when. Under his leadership, the DOE solar program would continue to be very supportive of small business participation; insisting that funds allocated to field operations be subcontracted out, San Martin expected and demanded a significant increase in contracting and subcontracting for small businesses.

5.3 The Solar Program's Field Support Structure

The implementation of the solar energy programs for R&D, demonstration, and commercialization involved the participation of a variety of organizations (as discussed in chapters 3 and 4, this volume). They included other government agencies, government laboratories, universities, and private companies of all types (see figure 1.9). In the early years of the program, NSF implemented its programs primarily through funding proposals from universities and individuals to conduct specific research; it tried to achieve a broad distribution by funding many small research efforts. In addition to grants to universities and individual researchers, the NSF/RANN program also funded some small programs at other agencies and government laboratories such as AEC's Los Alamos Scientific Laboratory and the NASA Lewis Research Center.

When ERDA was formed in 1975, the field programs from NSF, AEC, DOI, and EPA were brought into ERDA's program. The research and development capability of AEC, in particular, was extensive. As the program expanded under ERDA, virtually all of the former AEC laboratories became involved in solar energy R&D. At first, the laboratory involvement was restricted to in-house R&D, but as the program grew, the laboratories and field offices were asked to take part in the planning and management of contractor projects in their areas of expertise as well. By the time DOE was formed, there was a fairly formal management structure that included both the laboratories and the DOE field offices in the management of various aspects of the program.

5.3.1 The Roles of the National Laboratories

Each of the major solar technology programs set up field organization structures that best suited the needs of their programs. The Office of Solar Applications for Buildings, which the author headed, set up a broad support structure based on the requirements of its RD&D plans (*ERDA-77-144*). It tried to match the needs of the research program with the strengths and interests of the national laboratories and other research institutions. Because the solar building program was diverse and did not require large or unusual facilitates or resources, the supporting structure also tended to be diverse, with many of the laboratories and field offices participating. Over the years, the composition of this team and the percentage of the budget that each received would change, SERI would be established, and the RSECs would be added, then phased out, leaving only SERI.

On the other hand, because the solar power program involved relatively few, major engineering and test projects, its field efforts were concentrated in a few of the larger national laboratories that were in favorable locations for solar development and that possessed large and effective engineering resources. These projects were harder to move and tended to be more stable parts of the solar power program, even though their budgets underwent substantial variations.

In 1978 support roles for the solar building program were divided among a number of the national laboratories. Brookhaven National Laboratory (BNL) and Argonne National Laboratory (ANL) had a management support role for solar cooling and heat pumps and for solar energy storage, respectively; Lawrence Berkeley Laboratory (LBL) was involved in the management of active system controls and instrumentation and of passive cooling work; Los Alamos Scientific Laboratory (LASL, which later became Los Alamos National Laboratory or LANL) had management responsibilities for solar collectors, materials research, and passive solar heating; Lawrence Livermore Laboratory (LLL) assisted in the management of agricultural industrial process heating projects; and Sandia Laboratories which later became Sandia National

Laboratory Albuquerque or SNLA) supported advanced system research and systems analysis activities. At the same time, Sandia Laboratories, LLL, and JPL all had major R&D programs for solar thermal electric and industrial applications. Sandia had both in-house and contractor programs in line-focus collector technology and was involved in the central receiver program, LLL was the primary laboratory for central receiver R&D, and JPL was the primary laboratory for point focus (parabolic dish) technology as well as photovoltaics R&D. Three former AEC field offices, Albuquerque, Chicago, and San Francisco were also heavily involved in the management and distribution of funds for the R&D programs; the Chicago field office also had management oversight for the Commercial Buildings Demonstration Program.

In addition to the DOE laboratories and field offices, the solar programs were also implemented through agreements with the Department of Commerce (DOC), HUD, and NASA. DOC's National Bureau of Standards (NBS) conducted research and led activities to establish codes and standards for solar collectors and materials, solar water heating systems, and systems measurements of all kinds; it also had the unenviable assignment of evaluating unsolicited proposals from backyard inventors, many of whom claimed fantastic results for their inventions. HUD administered the residential solar heating and cooling demonstration program. In support of the DOE program, NASA's Marshall Space Flight Center continued some in-house R&D activities, and by 1978, had also assumed responsibility for administering the Commercial Solar Heating and Cooling Demonstration and the Solar Federal Buildings Program.

The DOE laboratories, NASA, and NBS provided the solar programs with a formidable set of assets: highly qualified senior engineers and scientists, strongly committed to solving the nation's energy problems and assuring national energy security, as well as the finest research facilities and the greatest concentration of computing power available anywhere. There were of course some management challenges associated with working with the established laboratories. Each laboratory group had its own "corporate culture," which was often very different from that of a particular DOE office and very hard to change. Most of the engineers or scientists, although not at first "solar" experts, came from programs that involved all the same basic elements of science as solar energy research and were accustomed to large interdisciplinary projects. The laboratories themselves viewed their new solar energy programs with mixed feelings: they were generally pleased with additional funds and with being involved in programs that were immensely popular with the public, but some senior managers resented the loss of talent from their mainstream programs, and some even felt solar energy R&D was beneath their "high-tech" dignity.

Los Alamos Scientific Laboratory was one of those labs where strong ambivalent feelings about the solar research program were most conspicuous. The laboratory's director in 1978, Harold Agnew, was a supporter of the LASL solar energy research program, but Associate Director Richard F. Taschek considered the program "low technology." (LASL solar program staff joked that Taschek's definition of low-tech was any technology where cost was a design criterion.) Taschek and others apparently viewed solar energy research as appropriate for backyard mechanics but not for Ph.D. physicists and engineers (eight of the twelve staff members in the LASL solar program at the time were Ph.D.s). They were especially disturbed by the passive solar work in which the objects of test and analysis were nothing more than ordinary houses. At one time, the director of the Energy Division of LASL actually called in the program's group leader and asked him to consider more esoteric approaches to solar energy utilization. "Couldn't you do it with lasers?" he asked.

When Director Agnew resigned in the fall of 1978, the interim management seized the opportunity to rid themselves of this embarrassment. The solar program was unilaterally removed from the LASL Institutional Plan issued in April of 1979. The excuse offered was that there were manpower constraints and pressure from DOD to reevaluate priorities. It was noted, however, that with nearly 7,000 employees and a great many nondefense projects, management chose to eliminate only the 27-person solar energy program.

Strong protests from Assistant Secretary for Conservation and Solar Applications (ASCS) Omi Walden were rejected by the interim management. Fortunately, Donald Kerr, former Acting Assistant Secretary for Energy Technology in DOE, was appointed director in the summer of 1979. Kerr reviewed the requests from the ASCS to continue the solar work and J. Douglas Balcomb's letter of resignation (in the event of program termination), and decided to retain the solar work. It was announced that Energy Under Secretary John Deutch, had agreed to raise the LASL ceiling from 6,800 to 6,850 persons to accommodate the continuation of the program.

The incident received a lot of attention from the local press in New Mexico, where solar energy was and is a very popular energy form. The Santa Fe *New Mexican* noted that Sandia's solar energy program was not threatened, presumably because they used lasers to measure the errors in the curvature of parabolic mirrors and must, therefore, be involved in "high-tech" research.

5.3.2 SERI Joins the Field Support Team

The creation of SERI provided the opportunity for a reexamination of the roles of the various laboratories and brought to the surface some of the issues that had been suppressed. It also brought to the fore the interlaboratory rivalries. In the "pre-SERI" program, each of the institutions involved in the solar heat technologies program negotiated a role for itself based on the match between its particular interests and expertise and the needs of the program. The program was large enough for everyone, and growing, and there was little rivalry. When SERI entered the picture, it had no established base of expertise and no research facilities, but it had the mission to conduct a broad range of solar R&D, and it was born with good connections at DOE, and the Congress.

At first, because SERI was busy hiring staff and organizing and reorganizing itself, there was little competition between it and the established laboratory programs. With the notable exception of the photovoltaics program, much of SERI's staff were young solar enthusiasts who did not have established research credentials. Moreover, two years after it inception, SERI was directed to take the lead in the commercialization of solar energy technology, a role that none of the other laboratories coveted.

After its controversial birth in March 1977 (discussed briefly in chapter 4, this volume), SERI began formal operation in July 1977, with Paul Rappaport as its first director and Michael Noland as its first deputy director. Rappaport, a prominent researcher in the field of photovoltaic devices at RCA Laboratories, had been the key to awarding SERI to Midwest Research Institute (MRI), SERI's parent company. By September 1977, just three months after SERI's formal start-up, Rappaport had filled the top positions in its five operating divisions. SERI began its operations in 30,000 square feet of rented space, expanded to 55,000 square feet in the second year, and to 90,000 square feet in the third year.

The initial staff of 72 people consisted of 43 professionals (researchers, analysts, and information and education specialists), 25 administrators, and 4 managers in the director's office.

By October 1978, SERI's staff numbered 370, representing the largest concentration of scientists and engineers ever assembled to work exclusively on the development of solar energy. Although management was planning to almost double the staff by the end of SERI's second year, to 700 professionals and support personnel, SERI's projected operating budget of $25 million in FY 1979 was only about half the size originally envisioned for SERI by NAS in its 1975 report.

When SERI did stake out its research territory, it placed among its first priorities the advancement of photovoltaics research and materials science. These two areas best fit the talent that Paul Rappaport newly hired or brought with him from RCA. SERI's first research projects in solar heating technologies were usually developed at the direction of DOE rather than self-initiated. Except for Frank Kreith, who initially retained his professorship at the University of Colorado while working part-time with SERI, the institute was not able to attract any "name" researchers or professors to work in its solar heat technologies program and often had to rely on contract research to accomplish the tasks assigned it by DOE.

During its first year of operation, SERI broadened its research priorities to include photovoltaic systems, biological and chemical energy conversion, wind energy conversion, passive solar energy techniques, energy storage, industrial process heat, and centralized and decentralized solar systems. Programs were begun in basic and applied research, economic and social science research, and planning and analysis. As mandated by Congress in the public law establishing SERI, a Solar Energy Information Data Bank (SEIDB) was created and began to work with the already existing National Solar Heating and Cooling Information Center. Other activities included participating in the Domestic Policy Review of solar energy and publishing its first annual review of solar energy. All in all, an impressive first year.

But all was not sweetness and light for the new SERI. Trouble was brewing in Congress over the potentially conflicting roles of SERI and the RSECs. After months of feuding fueled by RSEC supporters, principally from the Massachusetts delegation, Senator Floyd K. Haskell (D-CO), demanded that Secretary Schlesinger resolve the issue. Under Secretary Myers, in a letter to Senator Haskell dated March 10, 1978, spelled out the responsibilities and charters for SERI and the RSECs: "National SERI shall function as DOE's lead institution with regard to solar research, development and demonstration activities," while the RSECs, "in carrying out their primary mission ... shall be responsible, within their respective regions, as DOE's lead institutions related to the commercialization of solar technologies and conservation." This assignment of responsibilities calmed down the Colorado delegation and, to some degree, mollified the RSEC supporters because the RSECs had been given some specific, lead responsibilities.

This rather bitter debate was caused, to a large extent, by the complications mentioned earlier (chapter 4, this volume), when the establishment of DOE split solar programs between two assistant secretaries. Under this division, SERI oversight was assigned to the Office of Energy Technology, while RSEC oversight went to the Office of Conservation and Solar Applications. This division was soon to be eliminated, but in the meantime, bureaucratic and congressional rivalries kept the issue alive, even after the Myers solution. More than one year later, in October 1979, Denis Hayes, the new director of SERI, stated in testimony before the House Subcommittee on Energy, Development and Applications, "This division of accountability and programmatic guidance was one significant factor limiting the degree of cooperation which was possible between SERI and the Regional [Solar Energy] Centers. However, much of the responsibility rests squarely upon SERI and the regions, and reflects our collective unwillingness to decide which functions were appropriate for our respective organizations." Inside Washington and out, political feuds die hard.

In May 1979, in response to DOE's request that SERI support the aggressive commercialization of solar technologies, SERI changed its organizational structure. The new structure was a clear departure from its initial research orientation and was designed to provide the necessary framework to enable SERI to take a leadership role in the aggressive technology and market development of solar energy. The market development aspect of this new structure resulted in SERI being in direct competition with the RSECs (see sections 5.3.3 and 5.3.4).

SERI established two new technical divisions: Technology Development, headed by Charles Grosskreutz, and Technology Dissemination, headed by George Warfield. An MRI/SERI coordinating board was formed to review and monitor SERI's annual operating and long-range plans. Senior SERI management argued that a successful solar program depended on proper coordination of government work with private industry efforts. SERI's plans called for activities that would reduce the risk and uncertainty faced by the private sector in making investments in solar and by consumers in choosing between solar and other competing technologies and products. SERI management also noted that government regulations hampered solar commercialization: there were restrictions on paid advertising (based on the prohibition against lobbying with government funds), on government printing and technical information distribution, and on the ability of outside groups to meet with and provide advice and comments to government officials.

As part of its new commercialization and public education programs, SERI staff solicited information from community-based solar energy and conservation organizations to develop materials for other grassroots groups interested in conducting solar energy and energy conservation programs for community audiences.

In July 1979, in a surprise move, President Carter appointed Denis Hayes as director of SERI, replacing Paul Rappaport, who was in poor health. (Rappaport, caught by surprise, believed he was fired because he had been too independent and had lost support in the Colorado congressional delegation.) Hayes was the driving force behind the first Earth Day and a very vocal activist for solar energy. His appointment was supported by a large number of grassroots organizations that had formed to promote the greater use of solar energy, most notably, the Solar Lobby. The appointment of Denis Hayes was seen by many as an attempt to appease the solar activists and demonstrate the administration's commitment to solar energy. Hayes said that he intended "to help make SERI the most intellectually exciting, scientifically rigorous, and fundamentally honest institution in the world ... one whose integrity is beyond question." Hayes's appointment was welcomed by many solar proponents, who respected his oratory and leadership skills. The solar research community, however, was concerned about the wisdom of having a research institute led by a political activist.

In August 1979, Hayes named Henry Kelly, then the director of OTA's international programs, to be Assistant Director for Analysis at SERI. Kelly had been a frequent congressional witness on solar legislation and technology assessments, and was the principal author of OTA's two-volume report *Application of Solar Technology to Today's Energy Needs*, released in 1978. At SERI, Kelly was responsible for analysis of systems, policy, economic, and institutional questions, environmental assessment, and program planning. John Veigel was promoted to Assistant Director for Technology Commercialization and was responsible for promoting the use of solar energy technologies to decision makers in the consumer, commercial, utility, and industrial sectors, and in various levels of government.

5.3.3 The Regional Solar Energy Centers

In July 1977, at the beginning of his administration, President Carter met with most of the state governors and suggested a greater role for the states in energy matters. Carter said he would favor granting state and local officials more say in deciding such crucial issues as increased use of solar energy, the proper role for nuclear power, and the possibility of emergency allocations of scarce energy resources. However, he declined to say how he intended to take action on the states' recommendations. The president sought state involvement in advising Secretary Schlesinger on the need for incentives to reward a state for making investments in energy conservation, on the mechanism and structure of the soon-to-be-established DOE, and on the means for developing alternative energy sources.

Carter's energy policy envisioned a vigorous solar energy program designed to match the renewable energy resources of each region of the country to its energy needs. As mentioned earlier (chapter 4, this volume), before the establishment of DOE, the Carter administration had decided to allow the teams whose SERI proposals were unsuccessful to bid for planning grants to establish several regional solar energy centers (RSECs). Under loose solicitation guidelines set by ERDA, interested states were encouraged to organize into regional centers in the Northeast, North Central, West, and Southeast. The regional centers could decide to be anything from information-gathering and dissemination points to sophisticated research centers, whose work would be coordinated by SERI. In September 1977, DOE provided planning grants to two regional centers. The Midwest Regional Solar Energy Planning Venture, in Eagan, Minnesota, received $799,326; and the Western Regional Group received $697,500. Previously, ERDA had awarded a planning grant to the Northern Energy Corporation, in Cambridge, Massachusetts, for planning a regional center in New England, and DOE anticipated making a similar grant to the southeastern regional team.

Not unexpectedly, SERI was concerned that the anticipated high funding requirements of the four RSECs would cut into the $12.7 million total budget expected for SERI activities in FY 1978. Of even greater concern to SERI management was the future distribution of funds and authority. A DOE Management Operations Study Task Force that included Assistant Secretaries Donald Beattie and Robert Thorne reviewed the SERI and regional centers programs and made several recommendations to Secretary Schlesinger. It was decided that DOE and not SERI should oversee the activities of the regional solar energy centers.

The RSECs were chartered as nonprofit organizations, whose funding and work were initially approved by acting Assistant Secretary Beattie. When Omi Walden became the Assistant Secretary, she assumed that responsibility but then delegated management oversight to the director of the Office of Solar Applications. A board of directors was appointed to provide oversight of the management and programs of each of the RSECs. The governor of each participating state, with the concurrence of the Assistant Secretary for Conservation and Solar Applications, appointed one board member. DOE required a five-year contractual agreement with each of the centers.

DOE established somewhat loose guidelines for the RSECs and their roles and working relationships with the states and with SERI. DOE saw the need for the meaningful involvement of different groups, such as environmental and consumer groups in the RSEC programs. DOE believed the centers should pool the resources available in their states to augment DOE's efforts to accelerate the commercialization of the active, passive, and photovoltaic technologies, and that the states should have an active role in determining the use and distribution of the RSECs' funds.

The RSECs were required to maximize and develop whatever local resources were available, to encourage states to participate in their commercialization activities, and to draw on their expertise. Up to 10 percent of the RSEC's funds could be transferred to the states and additional money could be provided to the states on a project-by-project basis. The remainder of the funds were used to operate the RSECs and to support their various activities.

In January 1979, the newly established regional solar energy centers began work on solar outreach activities. The four RSECs were expected to coordinate state participation in solar energy and conservation programs, as well as help commercialize the technologies selected by DOE

Deputy Under Secretary Jackson Gouraud. The initial emphasis of the RSECs was placed on solar water heating systems and on passive solar applications. (Chapter 19 of volume 10 in this series describes the work done by the RSECs.)

5.3.4 Management Problems with SERI and the RSECs

In October 1979, soon after Assistant Secretary Stelson's appointment, witnesses testified before the House Science and Technology Subcommittee on Energy Development and Applications that SERI was an underutilized resource and that the RSECs seemed unclear about their mission. SERI Director Denis Hayes, one of the most critical witnesses, felt that SERI, rather than DOE, should be allowed to manage the RSECs on a task-by-task basis. While SERI and the RSECs were kept basically at arm's length, SERI's own chain of command required it to respond to four assistant secretaries and approximately fifteen senior managers at DOE. Because of perceived DOE restrictions, SERI had not been as aggressive as it might have been in disseminating information. Hayes also felt that SERI could make greater contributions internationally if it faced less bureaucratic red tape when it sought approval to initiate international solar activities.

Other critics also complained that there was a lack of direction from DOE for the RSECs. At first the RSECs claimed they were told they could structure their programs in any way they wanted. Later, after considerable effort had already been expended in designing R&D programs, they were informed that they must be involved in commercialization.

The GAO report *Solar Energy Research Institute and Regional Solar Energy Centers* (EMD-80-106), released August 18, 1980, claimed that SERI and the RSECs had neither been integrated effectively into the federal solar energy program nor achieved their intended objectives. The report noted that although DOE had reorganized its solar program structure to improve management and the use of SERI, it needed to do more. GAO suggestions included using SERI and the RSECs as lead institutions for solar energy development and commercialization; improving the planning process for developing the institute's and the centers' programs and activities; and monitoring the effectiveness of DOE's newly reorganized solar program, specifically with regard to the integration of SERI and the RSECs into the federal solar program and their use as lead institutions.

Earlier reports from the GAO had questioned the need, structure, function, and location of SERI and had examined the roles of SERI and the RSECs then under development. The new report pointed to conflicts over the roles of SERI and the RSECs in solar commercialization, saying that while DOE's charter statement assigned the RSECs a specific role in commercialization, SERI's role in that regard was not clear. RSEC staff viewed SERI's commercialization activities as an infringement on their responsibilities. SERI officials defended their programs by pointing out that most of the activities in question were initiated before the RSECs were placed under DOE contract; they were quick to criticize the RSECs for becoming too involved in the work of a national rather than a regional nature. The report noted, however, that SERI and the RSECs were actively working to improve their relationship.

Even before this GAO report had been issued, DOE initiated a series of meetings between senior SERI and RSEC managers to clarify the roles and responsibilities of these organizations. The basic distinction upheld by these meetings was that SERI's major focus should be R&D while the RSECs' should be commercialization.

5.4 The Evolution of the Solar Energy Industry Association

SEIA, the solar industry trade organization, was formed in the early 1970s to support the collective interests of the businesses in its field. It had its roots among the companies involved in a multiclient study of solar heating and cooling conducted immediately following the 1973 oil embargo by the consulting firm of Arthur D. Little. A specific proposal for organization of a trade association was put forward in 1973 by James Ince, the operator of a small management firm, and by Sam Taylor, a part-time member of Senator Hubert Humphrey's staff; subsequent discussions led to the formal organization of SEIA in the spring of 1974.

5.4.1 SEIA Support of Tax Credits

Sheldon Butt, SEIA's first president, made the association's first major objective to obtain the passage of solar energy incentive legislation, and to that end, much of the association's resources were devoted to lobbying activities and to the recurrent presentation of testimony to various congressional committees. The history of solar energy tax credits is largely the history of SEIA and its efforts to assure and retain the tax credit legislation. Sheldon Butt led the efforts to enact the legislation, and Jackson Gouraud, who was elected president of SEIA in 1983, led the fight to retain the tax credits that finally ended on December 31, 1995.

The first industry call for incentive legislation came in February 1974, even before SEIA was formally established, in a letter from Sheldon Butt to Senator Humphrey responding to a request for comments on the pending National Energy Research Act of 1974. In his testimony before the Senate Interior and Commerce Committees, Butt spelled out in some detail the industry's rationale for tax credits, setting the stage for SEIA's struggle to enact and preserve tax credits that continues to this day. Although eventually the industry succeeded in obtaining incentive legislation for some applications of solar energy, according to Butt, the legislation fell short of the industry's recommendations. SEIA had proposed a two-tiered residential tax credit structure to emphasize solar water heating and other low-cost investments, including solar swimming pool heating, and had recommended a much higher business tax credit. After passage of the original legislation in 1978, PL 95-618, the Energy Tax Act, SEIA continued it efforts to correct deficiencies in the bills as passed, but without success.

Throughout 1979, SEIA urged DOE to continue and expand support for tax credits and other incentives, development of standards, codes, and consumer protection mechanisms, and substantial media campaigns designed to educate the public. Butt outlined the solar initiatives he wanted the government to consider. These included a guaranteed low-cost loan program for homeowners and a bonus tax credit of 25 percent for solar retrofits of residential units to replace systems using oil, electricity, and natural gas. Butt proposed that investment tax credits be increased to 50 percent, adding that a guaranteed low-cost loan program was also needed for business, and that DOE's spending priorities be reordered. Half of the FY 1980 budget should be for commercialization, spending should be increased for industrial and agricultural applications and for solar cooling, and funding for systems development demonstrations and the Solar Federal Buildings Program should be terminated.

By the end of 1979, almost two years after his initial meeting with President Carter, Sheldon Butt charged that the administration had garnered attention and political advantage by appearing to support and even promote solar energy while actually doing very little to realize the president's goal that 20 percent of the nation's energy be supplied by solar energy by the year 2000. To achieve that goal, Butt argued, as he had for years, DOE needed to develop a dynamic, nationwide program to educate the public; to provide additional funding for voluntary testing; the Solar Bank needed to offer loan guarantees, not just interest subsidies; and Congress needed to raise the commercial and residential tax credits by 50 percent and to grant passive solar tax credits to builders. Butt called for expanding solar tax credits to include leasing of solar equipment, for international market development, and for greater funding of state and local solar efforts. Reversing his position on the Solar Federal Buildings Program, Butt supported its continuation but urged simplified regulations and a faster implementation schedule.

5.4.2 SEIA and the Development of Solar Codes and Standards

By early 1978, SEIA had grown to eight divisions, more than fifteen standing committees, and four state chapters. It counted among its accomplishments the adoption of some federal and state tax credits, increased government programs for federal procurement of solar technologies, and the initiation of federally funded demonstration programs. Through its voluntary technical committees, it had also initiated the development of a national testing, certification, rating, and labeling program for solar equipment. The efforts were financially supported by the Energy Information Administration (EIA) and DOE and relied heavily on the technical expertise of the National Bureau of Standards and other national laboratories.

SEIA, which believed from the beginning that consumers or end users represented its members' primary market for solar heating and cooling technologies, realized that it would have to convince consumers that it was equally interested in consumer protection. A major factor in accomplishing this objective was the development of solar equipment codes and standards through cooperative efforts with consumer groups and with state and federal governments.

SEIA's product certification efforts were mainly conducted through the Solar Energy Research and Education Foundation (SEREF), founded by SEIA in 1977 primarily for this purpose. Under a contract from the EIA, SEREF soon began work on a national rating, certification, and labeling (RCL) program through its Solar Standards Steering and Oversight Committee, composed of representatives of the professional societies, the utility, insurance, and banking sectors, consumer organizations, architectural and design firms, and other trade organizations. Most of its members were not affiliated with SEIA or SEIA members. Because the SEREF RCL program was broad-based and in the public domain by virtue of its government funding, in December 1979 SEIA embarked on a parallel RCL program for solar collectors to assure that the interests of it members would not be compromised. Approved solar collector models would have a SEIA label indicating their all-day thermal performance in Btus. The SEIA RCL effort, however, adopted procedures that were largely based on the SEREF approaches; the few differences in methodology were in nonsubstantive technical areas.

There were other organizations working on codes and standards at the same time. The Air-Conditioning and Refrigeration Institute (ARI) developed an independent approach for RCL, which was also based largely on the work done at NBS, and several states developed their own procedures and requirements. The federal government also was interested in developing a government RCL program similar to the Energy Efficiency Ratings (EER) adopted for energy conservation equipment. In early 1979, concerned about the proliferation of incompatible features of various standards and certification programs around the country, representatives of Florida and California held a meeting with DOE to discuss the concept of a national RCL program. DOE agreed to support the development of a single, consensus program and the Interstate Solar Coordinating Committee (ISCC) was created to achieve that objective. At about the same time, SEIA/SEREF and ARI attempted to combine their respective RCL programs in order to forestall government-imposed standards such as EER for solar manufacturers. Although the differences between the two programs were few, the organizations were ultimately unable to resolve them, and ARI went its separate way.

In order to retain a private sector RCL program, SEIA and ISCC agreed to form a new, independent certification organization to provide fee-based rating services to the solar industry. The Solar Ratings and Certification Corporation (SRCC) was incorporated in Washington, D.C., in 1980. To assure fair representation of both the solar industry and the public interest, its initial board of directors was drawn from SEIA and ISCC. After the downturn in the solar industry following the expiration of the solar tax credits in 1985, the SRCC is the only rating service that today provides rating, certification, and labeling for solar collectors and solar water heating systems. The SRCC currently operates three major solar programs: collector certification (OG-100), water heating system certification (OG-200), and a preferred optional rating and certification for solar water heating systems (OG-300). The OG-300 program integrates results of collector tests and system tests and determines whether a system meets minimum standards for system durability, reliability, safety, and operation; as well as factors affecting total system design, installation, maintenance and service. (For a complete history of the evolution of solar heating technology codes and standards, as well as the details of standards and methodologies, see chapter 14 of volume 10 in this series; for details of most testing procedures, see also volume 5.)

In addition to its work in codes and standards, SEIA also sought to improve quality and consistency throughout the industry through various training programs. In April 1979, both to further the goal of ensuring installation quality standards and to create jobs, SEIA and the Job Corps launched a training program for solar installation and maintenance workers. SEIA conducted two 6-month training sessions for 120 participants. Upon completion, the trainees received job placement and referral services from a special SEIA unit: the Job Corps SEIA Advisory Council.

5.5 Major New Legislation

Although it was sometimes called the "do nothing Congress," the 95th Congress actually passed some legislation that was very important to the evolution of the solar energy programs and to the solar industry. Table 5.1 lists the bills passed by the 95th and 96th Congresses that influenced the solar heat technologies program within DOE. There were many other bills that contained some solar energy provisions that affected other departments, including DOD, HUD, and USDA. Certainly the most important bill from the DOE perspective was PL 95-91, which established DOE. From the solar industry perspective, the two tax laws, PL 95-618, which established the residential and commercial tax credits, and PL 96-223, which increased them, were the most important. (For detailed converage of the full spectrum of solar legislation, see chapter 2 of volume 10 in this series.)

5.6 Situation at the End of the Carter Administration

The elements that supported a continued, strong federal government solar policy were in place when President Ronald Reagan was elected, along

Table 5.1
Significant legislation passed by the 95th and 96th Congresses

Public Law no. and date passed	Title	Pertinent provisions for solar heat technologies
95-91 8/4/77	Department of Energy Organization Act	Establishes DOE, creates the position of Assistant Secretary for Solar Energy Research and Development, and transfers ERDA functions to DOE
95-238 2/25/78	Department of Energy Act of 1978—Civilian applications	Authorizes loans and guarantees for RD&D of solar thermal heating and cooling and "other" solar programs.
95-618 11/9/78	Energy Tax Act of 1978	Provides tax credits for residential and commercial wind and solar energy equipment.
95-619 11/9/78	National Energy Conservation Policy Act	Directs that solar heating and cooling be demonstrated in federal buildings.
96-223 4/2/80	Crude Oil Windfall Profit Tax Act of 1980	Increases tax credits of PL 95-618. Qualifies solar panels installed on roofs of residences for energy credit.

with the first Republican Senate majority in twenty-six years. These were a well-defined R&D program, numerous commercialization activities, an extensive field support structure, and the involvement of solar industry researchers across the country. The Senate committees, however, would no longer be dominated by Democrats who had supported a strong federal role in promoting solar energy technology.

In spite of the Carter administration's professed support for solar energy, the first deep cuts ever experienced in a DOE budget for renewable energy and conservation were made by his administration prior to leaving office. The proposed FY 1981 budget for solar was reduced from about $819 million to $540 million; the Carter OMB cut an additional $34 million, bringing the final request to $506 million.

Note

1. In October 1978, preliminary plans were approved for installation of solar water heating systems on the White House, the Rayburn, and the Annex buildings, and systems were being studied for installation on the Hart, Forrestal, and Old Post Office buildings. The water heating load of the White House was being met by a central steam system, which provided heat at a very low cost. It was obvious that the proposed solar water heating system for the White House, which was expected to provide about 76 percent of the hot water requirements of the staff kitchen, barbershop, restrooms, and shower, had value as a highly visible dem-

onstration of this solar technology beyond that of just heating water. At a savings rate of $1,000 a year, the value of the displaced steam, the solar water heating system was unlikely to save enough energy during its lifetime to recover its projected $24,000 cost. The system turned out to cost a total of $41,770, reflecting the overdesign and overpricing that plagued many federal installations in those days, but especially the unique requirements of a White House installation. Although performing well, the system was removed in 1986, when the roof was resurfaced. It was not reinstalled because the Reagan staff considered it not to be cost-effective to do so.

6 The Contraction and Redirection Years: 1981–1988 (and Beyond)

Frederick H. Morse

6.1 The First Reagan Administration

6.1.1 The Transition Team

In December 1980 Ronald Reagan's energy advisory transition team and the Heritage Foundation, through its report *Mandate for Leadership*, presented the new administration with a wide variety of recommendations dealing with energy policy and the activities of the Department of Energy. Both groups recommended, at the top of their lists, that DOE be abolished and its essential functions transferred to other departments. Important recommendations related to the solar energy programs were to eliminate government-funded commercialization activities and to discourage SERI from engaging in social activism. On the positive side was the recommendation to increase funding for basic research and perhaps return to an ERDA-like agency to oversee such work. In regards to renewable energy, (the term they preferred to use for solar energy), the transition team expressed the view that renewable energy resources had substantial potential to contribute to the nation's energy needs. However, they urged that caution be exercised in increasing funding for renewable energy. The underlying premise of the report was that the administration should place greater emphasis on market forces.

The transition team recommended that the federal government (1) perform a reliable cost/benefit analysis of various renewable energy options, (2) support research and development of technologies not yet fully commercialized, and (3) justify the funding of renewables R&D on predicted performance and returns from the technologies. The transition team's report stated that much government funding went to renewable energy technologies that were "patently inefficient and did not live up to promised performance." The report also stated that DOE had become too large and unmanageable. It also recommended that controls on oil prices be abolished and that the DOE-managed defense and weapons R&D be transferred to DOD.

John Busterud, who led the part of Reagan's energy transition team dealing with conservation and solar applications, felt that President Carter's goal of solar energy providing 20 percent of the nation's energy by the year 2000 was on the high side. Busterud said he would review

Carter's Domestic Policy Review and recent SERI reports in the course of establishing the Reagan administration's energy objectives. He believed that the deregulation of oil would stimulate the use of coal and thus make solar more attractive.

Busterud believed that the best role for any research program was long-range work that industry could not or would not perform on its own. He did not feel that commercialization was a good role for the federal government, nor that the government had received good results from its investment in SERI, and thought it should be better managed. Busterud's recommendations characterized the philosophy on government's role in energy RD&D that would prevail during the Reagan years in the White House.

The impact that this philosophy had on the active and passive solar energy programs was profound and mostly negative. For example, program activities carefully designed to support the emerging solar water heating industry were quickly terminated, leaving that industry without critically needed support. Projects on durability testing of collector materials, research supporting codes and standards revisions for solar water heaters, and performance monitoring were terminated. Thus the rate at which the active solar energy system technologies could be improved by the solar industry was certainly reduced.

6.1.2 The Appointment of James Edwards as DOE Secretary

Upon taking office in January 1981, Reagan appointed James Edwards, former governor of South Carolina, as Secretary of Energy. Edwards had been chairman of the Southern Governors Association and chairman of the National Governors Association's Nuclear Power Subcommittee. Perhaps influenced by the large nuclear fuel-reprocessing plant located in South Carolina, Edwards was pro-nuclear. However, as governor, he helped to create the South Carolina Energy Research Institute, which studied a variety of renewable energy technologies, and was also involved in the early development of the Southern Solar Energy Center (SSEC), one of the four regional solar energy centers. Having overseen the planning documents that were submitted to DOE when SSEC was created, Edwards was very supportive of SSEC during its formation, and came to his new position with a greater understanding of solar energy than his predecessors.

To no one's surprise, Edwards was an advocate of the free market, supported the deregulation of oil and natural gas, and was concerned about the long time required to license nuclear power plants. Edwards believed that the U.S. energy policy should focus on three issues: national security, the price of energy, and environmental impacts. For the first time, the environmental impacts of energy production and use were referenced with regard to U.S. policy, although the solar energy programs would not soon benefit from this connection. Edwards felt that the United States should develop all of its energy resources, including its renewable energy resources, but he backed off from President Carter's goal of having solar contribute 20 percent of the country's energy needs by the year 2000. In line with Busterud's 1980 recommendations and those of other Republican leaders, Edwards believed that the federal role was to support research, develop technology to the point where it "worked," and then leave the further development and commercialization to the private sector. He supported streamlining DOE and eliminating regulations, but keeping intact the nuclear, defense, and crude oil reserve programs.

6.1.3 The Attempt to Close the DOE

Following the Heritage Foundation and transition team recommendation, Reagan attempted to phase out DOE. The initial steps called for major management changes, which included reducing the number of assistant secretaries to four, but leaving the Assistant Secretary for Conservation and Renewables among the interim survivors.

Despite protests from many Democrats and Republicans, Reagan formally proposed that DOE be dismantled, that its core programs, including a scaled-down conservation and solar program, be assigned to a new, semiautonomous agency, the Energy Research and Technology Administration (ERTA), under the Secretary of Commerce, and that DOE's defense-related programs be assigned to DOD.

Almost immediately, however, Reagan ran into trouble with Congress over his plans to eliminate DOE. Surprisingly, his greatest opposition came in the Senate from the Republican-dominated committee structure that provided oversight to the DOE, DOI, and DOC. Chairman of the Senate Committee on Energy and Natural Resources James McClure (R-ID), who risked losing his position to Senator Robert Packwood (R-OR) if DOE were to be dismantled, was a major player in convincing the president not to continue pursuing his plan to abolish the Department of Energy.

While for politicians, the discussions and battles over whether or not to close the Department of Energy was part of their job, the impact on the morale of the program managers at DOE was very negative. Until the issue was put to rest, most program planning beyond a few months ceased, and the attention of the program offices focused on ways to protect their activities under a range of possible "closing options."

6.1.4 The New OMB Philosophy

Based on its pre- and postelection rhetoric, and its failed attempt to eliminate DOE, the Reagan administration was expected to make significant reductions in the solar energy budget. At the time of President Reagan's inauguration, DOE was operating under a continuing resolution; in February 1981, Reagan proposed massive cuts in the solar and other energy budgets. The OMB proposal for FY 1981 was to reduce the solar budget to $193 million. To justify this large a cut from the final Carter solar budget submission of $506 million, OMB Director David Stockman put forth the policy that federal funds should only support long-term R&D with high payoff potential. Over the eight years of the Reagan administration, this policy would cause the active and passive solar energy programs to terminate essentially all market development and support activities with the solar industry, both for active and passive solar products and designs. Whatever funds were appropriated would be for activities such as the development of better materials and improved understanding of system performance. Under this policy, the federal government would support solar technology development only up to the point where the technology was proven to be workable; the private sector would be left to perform the remaining R&D and then to commercialize these technologies. In Stockman's view, support for basic research, technology development of unproven solar energy systems, and limited data collection activities would continue, while support for demonstrations and commercialization would be eliminated.

It is instructive to take a look at the allocation of funds among the participants in the program support structure, and among the various private contractors who performed work as part of the program. An analysis of the R&D spending of the active and passive solar energy programs (now being called "solar building applications") conducted in 1982

Table 6.1
SEIA companies receiving the most funds from DOE for solar heat technology development, 1976–1982

Company	Funding ($ thousands)
General Electric	10,627
Honeywell	10,102
SAIC	2,105
Westinghouse	966
Solaron	882
Owens Illinois	834
Martin Marietta	531
Wyle Labs	427
Wormser	399
RCA	385

at SEIA's request for the period from 1976 through 1982, provides some interesting insight into the nature of the R&D program at that time.

During the 1976–1982 period, the total spending for all solar buildings applications was $557 million. This figure includes analysis, research, technology development, demonstration, information dissemination, training, market development, and other activities. The R&D spending was estimated at $245 million. Of that amount, $103 million (43%) went directly to contracts with large and small companies; the balance, $142 million (57%), was spent by the government laboratories, including SERI and the RSECs, universities, and architecture/engineering firms.

The ten companies receiving the greatest program support are shown in table 6.1. While SEIA members received only 28 percent of the private company funding, they received most of the Residential and Commercial Demonstration and the Solar Federal Building funds.

Early in the Reagan administration, the Office of Solar Heat Technologies called a meeting with the solar energy industry representatives to assess the potential impacts of the new OMB policy on the DOE program and the solar energy industry. After considerable discussion, the consensus was that (1) the active solar industry did not have sufficient profits to perform its own R&D and to commercialize newer technologies, and (2) the DOE program should terminate all technology and product development support for the industry and focus on the innovation needed for the future. Accepting the reality of the budget levels and restrictions, and believing that DOE should become the research resource for the industry, the active solar energy industry realized that it would have to remain with essentially the same products it presently had, until the administration's policies changed.

To further justify reducing the solar energy budget, Stockman also recommended that construction of SERI's permanent facility be delayed, until the mission and staffing requirements of the institute were "better" clarified, and that the Solar Bank be eliminated. The Reagan administration and Secretary of Energy James Edwards supported Stockman's views and agreed to go along with all of OMB's budget cuts, not only those affecting renewables. This was a marked change from previous administrations, in which the Secretary of Energy usually attempted to preserve programs by appealing for revisions to OMB's recommendations. However, the Democrat-controlled House Science and Technology Committee accepted neither the magnitude of the budget cut nor the underlying philosophy. And, to Stockman's surprise, after the administration's budget submission, neither did Secretary Edwards.

While Stockman expected all energy technologies to be subjected to the same pattern of fiscal restraint, Edwards, during budget committee markups, wanted to exempt the nuclear budget from general budget cutting and to retain the breeder reactor program, which had also been targeted for elimination. Edwards and the Congress prevailed, and the nuclear programs fared much better than solar energy programs throughout the Reagan administration.

6.1.5 Starving the Solar Program

In addition to cutting the solar budget, the Reagan administration also planned to repeal or undermine more than twenty-four authorization laws passed in previous years to promote solar heating and cooling technologies and a variety of supporting measures, such as the solar tax credits, the Solar Bank (which at that time did not seem to have much of a chance of ever opening), and PURPA, which, among other things, legislated that states require their electric utilities to purchase power from renewable energy power producers. Rather than fight the battles required to accomplish this attack on the authorization laws, the administration decided instead to push for major cutbacks in spending, which would keep the laws from being fully implemented. Ignoring the many laws passed to promote solar and other renewable energy technologies, and ignoring the many related tasks required of the government, OMB prepared a document called the "DOE-OMB Budget Cases," which would, if followed, shut down the objectionable parts of the federal solar energy program by starving them of funds. During the second Reagan administration, the OMB followed the underlying philosophy outlined in the "budget cases."

6.1.6 The Appointment of Joseph Tribble as Assistant Secretary for Conservation and Solar Energy

In May 1981 Reagan appointee Joseph Tribble was confirmed as Assistant Secretary for Conservation and Solar Energy. Tribble had previously been energy coordinator for the Union Camp Corporation in Savannah, Georgia, and a local Reagan campaign chairman.

Tribble saw a limited role for government in advancing the use of solar energy. Tribble considered himself to be, not an advocate for solar energy, but a manager or administrator, carrying out the policies set by the Secretary of Energy, the White House, and Congress. Regarding Carter's 20 percent goal for solar, Tribble felt that predictions of this kind were generally wrong, although he agreed that renewable energy sources should be an important and more viable part of the nation's energy mix.

Consistent with the administration's line, Tribble thought that industry and financial institutions would respond to the nation's needs for renewable energy technologies as these became economical. According to Tribble, the government's role was to assist the solar industry in areas in which R&D was difficult, either because the costs were too high or because the risks too great. He also believed that DOE could justify supporting the R&D necessary to reduce the costs of these obstacles and prove the technologies, at which point industry would take over its further development.

Tribble believed that there were already enough government-created incentives for solar, for example the tax credit for the installation of solar equipment and the solar energy tax credit for industry. He differed with the administration in thinking that PURPA had some good provisions, for example, removing institutional barriers to cogeneration. In the past, he noted, companies wishing to generate their own power came under the jurisdiction of the Public Utility Holding Company Act of 1935 and revisions, and were subject to oversight by their state's public utility commission, with the result that they chose not to develop cogeneration.

Tribble also believed that there should be no federally mandated Building Energy Performance Standards (BEPS) because these should be handled by private groups, such as ASHRAE or trade associations.

From shortly after he was appointed to the summer of 1982, Tribble conducted intensive program reviews and sought to eliminate all activities that were not clearly long-term, high-risk research. Whenever possible, the solar energy program managers did their best to accelerate the completion of the "politically incorrect" work rather than simply terminate it, as terminating often resulted in losing both the committed funds and the results. The program, which once ranged from applied research and technology development to industry support and market development, was narrowed, little by little, to one that was predominantly research.

In July 1982 Tribble listed his office's accomplishments under the Reagan administration. Included were the previous administration's start-up of the 10-megawatt (electric) solar thermal central receiver power plant in Barstow, California, increasing the sale of photovoltaic modules and domestic solar water heaters in the United States, reducing the budget for conservation and renewable energy from $1,818 million in FY 1981 to $1,016 million in FY 1982 and a proposed $360 million in FY 1983, reducing his office's staff from 691 to 416 between May 1981 and May 1982, and revising and reinstituting a management-by-objectives system for monitoring program activities. Other "accomplishments" attributed to Tribble's watch included closing the four RSECs and seeing that solar commercialization activities were "successfully" transferred to industry, with a resultant savings of approximately $20 million per year. (The transfer was, of course, imaginary; the solar industry had neither the profits nor the expertise to pick up the "transferred" commericalization activities.)

6.1.7 The Reduction in Force and Its Impacts

The large reduction in the workforce of DOE's Conservation and Renewable Energy Office matched the decline in spending. In March 1982 DOE was criticized by Congress for allowing staffing levels to decline to the point where fulfillment of congressionally mandated conservation and renewable energy programs was at risk. When House Conservation and Power Subcommittee Chairman Richard Ottinger (D-NY) demanded an explanation, Tribble asserted that proposed budget levels would lead to further reductions in force. The renewable energy staff was expected to be reduced from 140 to 59 by the end of FY 1982. When the reduction in force was completed, many of the most experienced managers in the renewables program had left the department; many of those who remained, in this author's view, were dropped into a bureaucratic version of a giant Mix Master and transferred to programs they knew nothing about. As a result, the remaining staff found themselves overworked and lacking many of the skills they needed to effectively conduct the program.

6.1.8 Changes at SERI and the Field Support Structure

As the solar energy program grew during the Carter administration, the field support structure grew commensurately. The national laboratories and DOE field operating offices all found appropriate roles in the growing program.

This all changed when President Reagan was inaugurated and his appointees set about closing down the RSECs, eliminating commercialization from the solar program, and dramatically reducing the budget for all solar energy R&D. Initially, because so much of its work was considered to be commercialization, SERI was hit the hardest, losing nearly half of its staff in the first year of the Reagan administration. The other DOE laboratories fared better at first because much of their work was considered "long-term, high-risk" research, which the administration claimed to support. But the shrinking budget quickly fell below the level required to sustain even the most acceptable of the solar energy program's research portfolio. At the same time, the political decision was made to retain SERI at a level of about 500 employees. Faced with the double requirement of reducing research outlays while retaining a prescribed level of effort at SERI, the DOE program managers had no other choice but to reduce or close down research work at the other DOE laboratories and transfer the responsibility for the work to SERI. In most instances, projects simply died and were either forgotten or reborn within organizations at SERI that often lacked the expertise or facilities to carry them out. Very few of the leading researchers or managers from the other laboratories followed their interests to SERI. Notable exceptions were J. Douglas Balcomb (editor of volume 7 in this series), who gave up a good position and career with Los Alamos National Laboratory to maintain his involvement in passive solar energy research at SERI, and Barry Butler, who left Sandia National Laboratory in Albuquerque to head SERI's materials research work.

By 1984, Argonne and Brookhaven National Laboratories and Lawrence Livermore Laboratory had dropped out of the solar buildings program and the agricultural and industrial process heat programs, and little was left of the solar groups at Los Alamos National Laboratory and Lawrence Berkeley Laboratory.

The solar thermal electric program transferred responsibility for the point focus collector program from JPL to Sandia National Laboratory Albuquerque (SNLA; formerly Sandia Laboratories) and responsibility for the central receiver program from Sandia National Laboratory Livermore (SNLL) to SNLA. SERI was given the responsibility for advanced concepts and applications, but the bulk of the development work remained at SNLA. All the photovoltaic work was eliminated at JPL and transferred to SERI, which strengthened the kernel of expertise initiated by Paul Rappaport and turned SERI into a viable photovoltaic research center.

In keeping with the Reagan administration's lowering of the national priority of energy R&D, the new senior management at NASA, James Beggs and Hans Mark, also began to terminate NASA's energy work. By this time, NASA was heavily involved in many different types of energy R&D, approaching $300 million in work for other agencies, most in the field of renewable energy. The decision to terminate this work had the effect of significantly reducing the technical and managerial skills available to DOE and other agencies involved in funding energy RD&D, not to mention contributing to the confusion and uncertainty over how to continue and where to place hundreds of projects. To accommodate this new direction, after ten years of involvement, the NASA centers at Lewis, Huntsville, and JPL began the difficult process of transferring or terminating programs. By 1983, this process was essentially complete and only a few programs, such as wind turbine development, remained under NASA management.

As previously stated, SERI and the RSECs were criticized by Congress during both the Carter and Reagan administrations for not having well-defined and distinctly separate missions. In spite of efforts by DOE to sort this out, there was still disagreement between SERI and the four RSECs over whether SERI should be conducting commercialization work. However, as the Reagan administration had made clear, the president did not want the federal government to support commercialization activities. Fighting between SERI and the RSECs over these activities made little sense, especially because the RSECs, as leaders in the DOE solar commercialization activities, were in danger of losing all of their funding.

With a staff of about 950 people, SERI had anticipated a budget of $124 million in FY 1981. However, given the new administration's position, it was clear that both staff and budget would soon decrease. In March 1981 SERI Director Denis Hayes sent a proposal to DOE for reorganizing the four-year-old institute, consistent with the reduced role envisioned for it and for DOE in general. In his proposal, managerial positions would be reduced, and a large number of employees terminated.

In an attempt to end SERI's commericalization work and to redirect its activities to research, like the other national energy laboratories, DOE in early 1981 sent a three-person team, headed by Frank DeGeorge, Acting Assistant Secretary for Conservation and Renewable Energy, to Golden, Colorado, to review each SERI activity and to place it in one of six categories, ranging from the most acceptable (long-term, high-risk research) to the least acceptable (commercialization) activities. The review concluded that while the amount of work it characterized as non-R&D and therefore undesirable varied considerably from technology to technology, half of SERI's activities could be redirected to more acceptable research categories. The review team recommended that approximately 40 percent of SERI's non-R&D activities be completed because the bulk of the funds had already been expended; about 10 percent of ongoing activities were to be terminated outright. The study identified $18 million in saving on non-R&D activities, and those funds were returned to headquarters.

Controversy over SERI had been building for some time. An outspoken proponent of solar, Director Hayes was sometimes an embarrassment to the Carter administration, and certainly to the Reagan administration; he was fired in June 1981, following heated exchanges with newly appointed Assistant Secretary Tribble over proposed cutbacks in staffing for the institute. DOE was contemplating a reduction of as many as 370 positions, to run in line with a proposed, much-reduced SERI budget of $50 million. (In fact, SERI's budget in FY 1982 was $48.7 million, and the reductions to the SERI staff exceeded 370.)

6.1.9 The Appointment of Harold Hubbard as SERI Director

In August 1981 Harold Hubbard, then senior vice president of Midwest Research Institute, was named the new director of SERI. Anticipating budgets as low as $40 million per year for the next five years, Hubbard continued to terminate commercialization activities and focus on conducting and coordinating long-term, high-risk research and development "which private industry cannot reasonably be expected to undertake." He told the House Science Subcommittee on Energy Development and Applications that he was "optimistic about SERI's future, based on the assumption that program continuity and our present level of effort will be sustained through the next five years." Hubbard further reduced SERI staff, to about 520 permanent employees; job losses were greatest in nontechnical areas such as information dissemination, socioeconomic studies, and international, public, and consumer education programs.

Hubbard believed that the program and staffing changes put SERI in a better position to carry out its original and fundamental mission of research and development. Under Hubbard's protective leadership and skillful management, SERI survived its greatest threat and emerged as a strong, first-class research facility, with a primary emphasis on photovoltaics; solar fuels and chemicals; and low- and intermediate-temperature thermal processes and materials.

6.1.10 A Short History of the RSECs

In all, the RSECs existed for less than four years; their tenure was marked by conflicts over their mission, their relationship to SERI, DOE, and their participating states, their own management, and, of course, their budgets. With much of their time spent in staffing up, organizing, and closing down, it is no surprise that the RSECs left few notable accomplishments. They did, however, engage in a variety of activities, and within their regions conducted some effective programs.

Information dissemination was the key to most of the RSECs commercialization work. The centers maintained information programs that provided access and regional input to the national Solar Energy Information Data Bank (SEIDB), disseminated solar codes and databases within their regions, handled numerous public inquiries, and conducted workshops and training programs. They also distributed various solar information publications, provided direct consultation on technical and business questions, organized and participated in energy fairs in their regions, and made speaking appearances and radio or TV appearances that reached broad audiences. The Mid-America Solar Energy Complex (MASEC) estimated that their information program reached 5.3 million people, or about 9 percent of the region's population.

The RSECs also conducted specific programs designed to promote the acceptance of active, passive, and photovoltaic solar applications, and some conducted or sponsored studies related to biomass or wind energy applications as well. MASEC, for example, developed Solar 80, a competitive design-and-build program for passive solar houses, to reduce the annual fossil fuel use in homes by 80 percent, compared to the regional average. Plans for nineteen of the homes evolving from Solar 80 were sold through regional publications and *Better Homes and Gardens*; in all, over 1,500 sets of plans were sold. In addition, 13,000 building professionals and members of the public attended Solar 80 workshops sponsored by MASEC.

MASEC also initiated a program with the National Association of Home Builders in which a large-volume home builder was to adapt the design of one of its standard home designs to include active solar heating features, build it, and compare the results from instrumented solar and nonsolar homes monitored for a year. This program was transferred to DOE when MASEC ceased operations in 1982. The Northeast Solar Energy Center (NESEC) concentrated on photovoltaic applications and established an information center that survived NESEC's closing and exists today. The Southern Solar Energy Center (SSEC) focused primarily on solar water heating applications and worked with state energy offices and utility companies to assess field performance and assure quality installations. And the Western Solar Utilization Network (Western SUN) worked within the framework of a Compact of States to coordinate all energy policy for the region.

Perhaps the greatest contribution of the RSECs was to bring a more diverse and local input into the process of promoting the use of solar energy. All the RSECs maintained close ties to their state energy offices and channeled a significant part of their funding through those offices. This provided a strong linkage between the state and national programs. In addition, as part of their planning process, some of the RSECs reached out to a diversified group of experts in all aspects of energy and commerce for programmatic guidance and input. MASEC, for example, had its Solar Resource Advisory Panel, which ultimately grew to 6,300 members from all walks of life, who contributed to planning through a Delphi-Knossos polling process. These outreach activities would have undoubtedly continued and expanded but for the new perspective put in place by a new administration.

As the Reagan administration's policy regarding the solar energy program and budget became clearer, the four RSECs realized that their closing was a real possibility. A temporary restraining order was issued in January 1982, prohibiting DOE from terminating its contract with the Northeast Solar Energy Center until a hearing could be held. The RSECs were allowed to operate, but they found it difficult to continue with their planned activities at the level of funding in the FY 1982 appropriations bill. The $2.5 to $3 million that was expected to be released to each RSEC to pay for salaries and the daily operations of the offices, did not leave much for awarding subcontracts.

DOE's efforts to close the four regional solar centers continued, and by March 1982, the fight was over; they had been given final orders to shut down. By the end of that month, most of their subcontracts were terminated, and the employees were given their layoff notices. The closing of the RSECs brought to a premature end an innovative approach for supporting the commercialization of solar energy technologies where federal funds were leveraged with state funds to reduce the market barriers to the greater use of renewable energy technologies. Thus ended one of the more innovative and diverse approaches to commercialization of solar technology before it really had a chance to prove its merit. The short, four-year life of these centers, however, precludes any realistic assessment of their impact.

6.1.11 The Solar Energy and Energy Conservation Bank

Since its first mention by President Carter in June 1979, the Solar Energy and Energy Conservation Bank (the Solar Bank) remained an idea struggling to become a reality, and most of the time more dead than alive. The Solar Bank was created by PL 96-294, the Energy Security Act, signed into law by President Carter on June 30, 1980. Carter's DPR of solar energy cited the need for billions of dollars in new capital to implement its base case scenario, and envisioned the Solar Bank as a means of encouraging the necessary cash flow and assisting borrowers with financing costs for solar technologies in the residential and commerical sectors. Even before the DPR, similar legislation had been introduced by a number of legislators including Rep. Stephen L. Neal (D-NC) and Rep. Stewart B. McKinney (R-CT). The final law incorporated elements of these bills, the administration's proposals, and key alterations introduced along the way through a series of amendments. The final bill was a hybrid of the different approaches. It incorporated many of the recommendations of the pro-solar and public interest communities. As enacted, it served more to provide subsidies for those who could not take advantage of the tax credits (low- and medium-income families), than a general mechanism for financing the conversion from conventional energy systems to renewable energy sources as envisioned in the DPR.

Granted an initial appropriation of $100 million, with its staff positions filled and its operating regulations drafted by the Carter administration, the Solar Bank had an ambitious agenda. In January 1981, however, the Reagan administration withdrew the bank's operating regulations, dismissed its staff, and impounded its funds, pending a budget rescission that Congress eventually approved. All the bank's FY 1981 funding was lost except for $250,000 of administrative costs.

In August 1981 Congress was on the brink of reviving the Solar Bank, despite the best efforts of the OMB and HUD to prevent that and despite the Reagan administration's insistence that the bank was an unwise and unnecessary way to promote the use of solar energy and energy conservation. Overriding the administration's position, Congress voted to appropriate $25 million for FY 1982, and added language for the bank to implement all of its activities. Specifically, Congress called for moving rapidly to publish regulations, staff the bank, and dispense loans and subsidies at the earliest possible date.

Over six months later, with the bill passed, the administration still refused to act. In April 1982, five congressmen, two cities, one state, and several small consumer groups and individuals filed suit against President Reagan and six members of his cabinet to force the administration to release the regulations and begin funding of the Solar Energy and Energy Conservation Bank. The suit, brought before the U.S. District Court for the Southern District of New York, named as defendants President Reagan, OMB Director David Stockman, HUD Secretary Samuel Pierce, DOE Secretary James Edwards, Treasury Secretary Donald Regan, USDA Secretary John Block, and Commerce Secretary Malcolm Baldridge. According to the Solar Lobby, it was the first time since the Nixon administration that a president had refused to faithfully execute legal spending requirements. The suit sought declaratory and injunctive relief against the defendants in their official capacities as members of the Board of Directors of the Solar Bank, to enjoin their refusal to spend funds specifically appropriated to operate the bank, to hire staff, and to issue regulations.

As a result of the suit, OMB Director Stockman approved the transfer of $21.85 million to HUD for the bank. However, not giving up easily, the administration subsequently asked Congress to rescind these funds and to transfer them to the low-income home energy conversion assistance program. When the 45-day deadline for the president's requested rescission of the bank's funding passed without congressional action, HUD directed that Bert Fulmer, controller for the Government National Mortgage Association (GNMA or "Ginnie Mae"), work with Ginnie Mae President Robert Karpe, manager of the Solar Bank, to start the bank's organization. Plans outlined by Fulmer called for the bank to be running by the end of FY 1982, when the appropriation called for 80 percent of the conservation loan money and 70 percent of the solar loan money to be issued. However, before any loans could be distributed, HUD had to formally issue the bank's lending regulations, drafted during the last days of President Carter's administration but immediately withdrawn by the incoming administration.

In July 1982 the New York district court ordered HUD to create the Solar Bank as soon as possible, but dismissed allegations that funds for the bank had been illegally impounded and declined to set a specific deadline for start-up. The first loans supported by the Solar Bank were expected to be issued in early 1983. Help from the federal government, however, was not expected to last longer than a single year, according to Ginnie Mae President Karpe. Karpe said that his goal was to get the money into the hands of the states that had existing programs and that could effectively use these funds. The proposed loan programs, a buy-down of interest or principal, were to be directed to lower- and middle-income people and would be issued for conservation and passive solar options. Karpe said he hoped these incentives would provide an inducement for people to produce these technologies. With slightly over $21 million for the first year, the bank would be able to reduce the interest rate on a large number of loans, but still far fewer than anticipated. Karpe hoped that the states would come in with other programs, to have the most cost-effective outcome and to be a significant benefit to as many as possible. But the decision was up to the states, which knew their energy needs better than the federal government. By January 1983, all of the states, U.S. territories, and the District of Columbia had been informed by HUD that they would receive a portion of the Solar Bank's funds.

The struggle between the administration and the Congress continued. Each year, the administration would propose no funds for the bank, and Congress, led by the House, would restore $20–25 million and insist that the bank be allowed to retain any unspent or returned funds. New funds were appropriated for the program after 1985.

By the end of 1988, the Solar Bank, administered by HUD, had disbursed a total of $71,400,000 to subsidize bank lending rates to customers seeking loans for the purchase of energy conservation or solar energy systems. Only $9.2 million was used for solar projects. (For additional details on the legislative background, litigation, administration action, operations, management, and accomplishments of the Solar Bank, see chapter 26 of volume 10 in this series.)

6.1.12 The Residential Conservation Service

The Residential Conservation Service (RCS) was authorized in PL 95-619, the National Energy Conservation Policy Act of 1978, as a way to promote the use of energy-efficient products. The objective of the RCS program was to provide advice and assistance to homeowners seeking to improve the energy efficiency of their houses; its implementation would bring major electric and gas utilities into the solar energy and conservation energy business by requiring them to perform conservation and solar audits of homes for prospective customers who might have a difficult time identifying their most cost-effective energy-saving measures.

The RCS program was first implemented by a few utilities in late 1980. In November 1981 the governors of the western states approved a resolution calling for continued availability of the solar tax credits because, in the absence of these credits, it would be more difficult for a homeowner to implement an energy audit's recommendations.

As a part of his plans for deregulation, President Reagan wanted the marketplace to continue to set the energy use patterns. In 1980 the Heritage Foundation's transition team report, *A Mandate for Leadership: Policy Management in a Conservative Administration*, had singled out the RCS program for sharp curtailment, in one of its 2,000 or more recommendations. Agreeing with the report, the administration considered the Residential Conservation Service to be an example of an intrusive program and thus one that should be discontinued. At the start of 1982, it appeared likely that the states would have to be responsible for saving the RCS from becoming a meaningless activity as far as solar home improvements were concerned. Solar lobbyists and consumer groups attacked DOE over the proposed changes in the RCS utility audit program for homeowners. Assistant Secretary Tribble, of course, favored the elimination of the RCS program; however, until the program could be legally discontinued or amended, DOE was obligated to carry out the law. The administration therefore revised existing program regulations to make them as simple and flexible as possible, consistent with existing law and sound management practices. DOE was prepared to further revise RCS's regulations enabling utilities to ignore solar options in most parts of the country when they performed the energy audits of ratepayers' houses required by the National Energy Conservation Policy Act of 1978. And during testimony at public hearings held in Washington, D.C., the utilities expressed relief that they might not have to conduct the home energy audits at all.

In the new, final regulations for the Residential Conservation Service announced by DOE in July 1982, DOE was to approve each state's implementation plans. The final regulations made a number of concessions to the solar industry, restoring aspects of the program that would tend to encourage consumers to use more renewable energy options and assuring them that they would be getting quality products if they chose to buy them based on the audit recommendations. Under the final rules, customers of any large regulated utility could have their homes audited for a wide variety of conservation and alternative energy retrofit options, provided the RCS plan adopted by their state determined the options were cost-effective. By summer 1982, thirty-six states had received final, and five conditional, approval of their RCS plans from DOE. As a result of the RCS program, energy audits were conducted in over 500,000 utility customer homes.

6.1.13 Tribble Departs

The political wrangling over the Solar Bank exemplified the relationship between Congress and the administration during Assistant Secretary Tribble's tenure. Much of Tribble's work at DOE was aimed toward downsizing his programs, in direct opposition to congressional mandates.

Committee members agreed that the marketplace ultimately had to decide the manner and degree to which renewable energy technologies would displace more traditional sources of energy, and that some basis had to be established to assist in determining the conditions under which federal R&D assistance would be provided. They did not agree, however, that most solar and renewables research had been brought to the point where the private sector was willing to take over on their own.

Consequently, the appropriations subcommittees recommended restoring program funds and staffing levels to carry out what it called the "basic research activity level." This recommendation would maintain programs and staff expertise in each of the solar and renewable program activities and would provide funds to continue related operations at SERI and other DOE research laboratories.

In a clear departure from the administration's wishes, funding was provided to complete several major demonstration projects and to collect data and information from ongoing programs so that the federal investment would not be lost. While the active and passive solar energy demonstration programs were already completed, this congressional recommendation provided moral support for the completion of the Solar Federal Buildings Program.

While the administration had not requested any money for the active solar heating and cooling program in the FY 1983 budget, Congress recommended that $6.65 million be appropriated for that program: $2 million for materials research; $1 million for integrated systems modeling; $500,000 for active cooling concepts; $500,000 for standards development; $75,000 for monitoring and data collection; $75,000 for technology transfer; and $2.5 million for other activities. Congress further recommended that $5 million be appropriated for the passive solar program: $1 million for materials research in glazing, thermal storage and insulation; $500,000 for systems integration; $500,000 for data collection on building thermal performance; $500,000 for technology transfer, and $2.5 million for other activities.

It should be mentioned that this level of detail, although common in appropriation language, had both positive and negative consequences, depending on the situation. Relating a dollar amount to a specific program element could serve to protect that activity should others wish to reduce or eliminate it. On the other hand, this level of detail often severely limited management's ability to optimize program content for a specific total budget.

6.1.14 The Appointment of Donald Hodel as DOE Secretary

In the autumn of 1982, as part of the consolidation of the Reagan administration's energy policy, Edwards stepped down from his position as Secretary of Energy and was replaced by Donald Hodel. Confirmed on December 8, 1982, Hodel was experienced in the energy field and had promoted the construction of five nuclear plants in Washington state during his tenure as administrator of the Bonneville Power Authority. During his confirmation hearings, Hodel promised a new DOE administration but was quick to point out he supported the president's energy priorities and policies. Hodel strongly believed the country needed a wide range of energy resources and technologies. In keeping with the president's focus, DOE would continue its involvement in "long-term, high-risk, high-payoff research and development." Although Hodel supported a stronger role for the federal government in renewable energy, he believed the most effective way to achieve real gains in conservation and the use of renewables was through the use of marketplace incentives, and favored extending the energy investment tax credits beyond 1985. Secretary Hodel wanted to keep the United States in the forefront of solar technological development, and he felt SERI was important to meeting that objective.

6.1.15 The Appointment of Pat Collins as Acting Assistant Secretary for Conservation and Renewable Energy

In August 1983 Pat Collins, former Vice President for Public Affairs at the National Association of Home Builders, was appointed Under Secretary of Energy. Shortly after Collins's confirmation, Joseph Tribble announced his resignation. A long-time proponent of energy conservation and solar energy, Collins was given a second job, to replace Joseph Tribble as the acting Assistant Secretary for Conservation and Renewable Energy. Collins, in a break with his predecessor, believed that the United States should promote the use of conservation and renewable energy technologies, and during his brief tenure defended a stronger program in the face of ever-declining budgets.

6.1.16 A New Energy Policy

In October 1983 DOE produced a formal document on energy policy titled *The National Energy Policy Plan*, which stated that the primary purpose of national energy policy was to promote a "balanced and mixed energy resource base," minimize federal control and involvement in energy markets, and maintain public health, safety, and environmental quality. "Balance" meant an economically efficient and flexible energy system in which the mix of energy supplies was appropriate to support economic growth and adequate to permit producers and consumers to choose freely among a range of energy options. Under this approach, it was expected that conservation would be viewed by policy makers, producers, and consumers as the equivalent of an important energy resource. This was perhaps the first time that DOE recognized conservation (soon to be called "energy efficiency") in this context, laying the foundation for the demand-side management (DSM) movement that was beginning at about that time. Renewable energy R&D would continue to focus on improving system efficiencies, reducing material costs, and increasing system life. In the solar buildings area, federal research was directed toward developing advanced materials that would reduce costs and improve performance of both active and passive systems.

According to the new DOE policy document, the United States had progressed from federal intervention to shield the economy from the realities of change brought about by energy crisis to confidence in the free market and the ability of consumers and businesses to ensure adequate energy supplies at reasonable costs. The market would promote economic efficiency through innovation, fuel substitution, and development of new technologies.

6.1.17 Reagan's First Term Comes to an End

As 1984 drew to a close, the outlook for government support of the development and commercialization of solar technologies was not good. Consumer spending for solar energy systems was down, and high interest rates restrained borrowing, especially for home buying and home improvement. The oil glut had lowered energy prices and diverted the public's interest in obtaining alternative energy sources; as a result, solar collector sales had slowed down. The Reagan administration, still hostile toward energy conservation and renewable energy, threatened to eliminate some or all of the energy tax credits and to further downgrade the Residential Conservation Service; it proposed a budget featuring a shrinking renewable energy program and an expanding nuclear energy program. However, some solar enthusiasts felt that solar could be at the start of a miniboom, provided the oil glut did not last and Congress protected solar tax credits, the renewable energy budget, and the Solar Bank. Such optimism proved short-lived. Worldwide oil prices remained low; the tax credits and the Solar Bank disappeared.

Energy was hardly an issue in the 1984 presidential campaign and neither party's platform made much of the energy needs of the country. But there were differences in the two parties' positions. Democrats called for increased support for conservation and solar energy, including renewal of the tax credits. The Republican platform again raised the issue of abolishing DOE and sought regulatory relief to permit bringing nuclear plants on line more quickly.

6.2 The Second Reagan Administration

Early in 1985 a House Republican research committee, in its report *Ideas for Tomorrow, Choices for Today*, called on the administration to curb direct government involvement in the commercialization of renewable energy technologies and to encourage the development and use of standardized nuclear power plant designs, while streamlining the licensing process. The report pointed out that although the government had spent billions of tax dollars to develop renewable energy sources, the results were disappointing and federal energy research was hampered by duplication of effort; the greatest benefit to applied energy research had been President Reagan's commitment to deregulation. It speculated that if voluntary financing were provided by the energy industry, politics could be further removed from research, and federal outlays could be reduced, although the government would continue to fund basic research. To privatize the R&D sector, scientific and educational organizations would issue tax-exempt bonds, the proceeds of which would finance facilities and equipment for colleges and universities. The report recommended a strict hands-off policy toward the development of renewable energy technologies by the federal government, and this recommendation, followed by the administration, set the stage for the next four years.

6.2.1 Refinements in OMB's Energy Philosophy

Early in Reagan's second administration, OMB expanded its rationale for ever-decreasing renewable energy budgets. In 1985 DOE officials outlined a plan to bring renewable energy products from the idea stage to the market, which called for joint expenditures by industry and DOE, with decreased government expenditures as the technology progressed toward the marketing stage, and increased responsibility falling on the firms that hoped to market the products. While the part about reducing government funding as products progressed along the development path was very similar to the approach introduced during Omi Walden's tenure and was well accepted by the DOE staff, the "joint ventures" part was new.

To make up for some of the relentless reductions in the federal solar energy budget, the new DOE plan envisioned a cooperative research and development venture pool to which several private solar energy companies would contribute about $55 million of their own funds to support broad R&D efforts and then share in the results equally. OMB went so far as to ask DOE to consider turning SERI into an industry cooperative R&D venture, or a R&D limited partnership with DOE as a minority partner or shareholder.

In a January 1986 letter addressed to the DOE Secretary, SEIA stated that OMB's funding strategy, wherein private companies would cofund federal energy R&D programs and share the results with each other, was unworkable for solar companies because they were too young and too small. According to SEIA, the solar industry lacked the experience and investment capabilities to support such a concept; over 90 percent of the industry had been in existence only since 1977. SEIA argued that solar energy companies and federal laboratories were functioning well together and making significant progress and that the relationship should remain unchanged. Furthermore, because the industry was relatively new, many firms had developed patented and proprietary items that needed to be protected.

In the fall of 1986 DOE announced that it would seek suggestions from renewable energy firms for cooperative R&D ventures. The venture structure called for groups of companies working on the same general topics to collaborate on long-range research of benefit to all, without getting into specific products where proprietary issues would arise. Funding would be evenly split between the private sector and the government. The government would not sponsor solar demonstration projects and would not accept construction projects under the cooperative R&D venture concept.

6.2.2 The Appointment of John S. Herrington as DOE Secretary

At the start of Reagan's second term, Donald Hodel was shifted from his position as Secretary of Energy to Secretary of the Interior, causing speculation that the two departments would be merged, as had been proposed in 1980 when the initial talk of closing down the Department of Energy began.

John S. Herrington, who had previously held senior positions at DOD and the White House and who had little energy experience, was nominated by Reagan in January 1985 to replace Hodel as Secretary of Energy. Herrington's priorities were very similar to Hodel's: natural gas deregulation, simplified nuclear plant licensing, energy tax policy, environmental protection and energy security. As he explained during his confirmation hearings, Herrington supported a federal role in developing solar energy and conservation, stressed the need for reliance on a balance of energy resources, although not necessarily in budget dollars, and favored tax incentives for energy investments. He did not support the closing of DOE.

Secretary Herrington believed that energy stability, security, and strength should be the objectives of U.S. energy policy. In his first major policy speech as Secretary of Energy, he focused on coal and nuclear energy but said that the United States had the potential "to turn increasingly to a full spectrum of renewable energy resources for clean, efficient energy."

6.2.3 The Appointment of Donna Fitzpatrick as Assistant Secretary for Conservation and Renewable Energy

March 1985 saw some significant changes in the leadership of the renewable and conservation programs at DOE. Pat Collins resigned as Under Secretary of Energy, significantly reducing the access of solar energy and energy conservation advocates to high-level attention at DOE. Donna Fitzpatrick, who had been Principal Deputy Assistant Secretary, was nominated as the Assistant Secretary for Conservation and Renewable Energy. During her confirmation hearings, Fitzpatrick explained that DOE's research work would continue but that its demonstration projects would not. (It is interesting to note that although Omi Walden began the process of phasing out the demonstration programs in 1979, seven years later they were still the subject of discussion and controversy.) Donna

Fitzpatrick said that DOE would base its decisions for trimming the solar program on three criteria: the potential for achieving lower costs; the chances for obtaining higher efficiencies; and the ability to achieve a higher reliability.

In May 1985, appearing before the Senate Energy Subcommittee, Fitzpatrick had difficulty convincing the senators that her proposed FY 1986 solar energy and conservation budget "strongly supports everything we should support," as she had testified. According to Fitzpatrick, the proposed budget preserved the core of the federal research effort, supporting important R&D activities while dropping spending for demonstration projects. However, with the president's tax simplification plan expected to call for the elimination of solar energy tax credits, Fitzpatrick admitted that there could be fewer companies to benefit from the federal R&D spending if Congress went along with the plan.

Fitzpatrick believed that during the second Reagan administration, DOE's emphasis would continue to be on research that was long-term and fairly high-risk; that the department would eliminate funding for relatively expensive demonstration projects but continue to support renewable research efforts at the same level as before. As for demonstration projects already underway, Fitzpatrick stressed her view that they were properly the responsibility of the private sector. The department would consider, on a case-by-case basis, participation in Proof of Concept Experiments (POCEs), subject to the availability of government resources, and sufficient resources to complete its responsibilities relating to technology demonstration projects begun in prior years.

Throughout the second Reagan administration, DOE management believed its investment should be focused on a broad-based research program, and that diverting funds to commercial demonstrations retarded the pace of the research program, froze technological development by the private sector, and created pressure for further subsidy of uneconomic or immature technologies.

During the three years since the reduction in force (RIF), the federal solar organization was consolidated from eight to five divisions and the entire branch structure was eliminated. When reviewed by the Energy Research Advisory Board Solar Panel in October 1985, this consolidation appeared to have resulted in marked efficiency improvements, particularly in management. In a letter to Under Secretary Joseph Salgado, the panel recommended "additional steps" to make the DOE renewables programs even more productive: continuing or increasing the level of university and industry activities; providing opportunities for collaboration, competition, creativity, and technology transfer, while assuring efficient and effective use of limited resources; continually reviewing programs to foster the winners; phasing out R&D as successful technologies were implemented; and terminating or redirecting funding for less promising technologies. Because the solar buildings programs were already doing all of these, the panel's recommendations had little programmatic impact; in this author's view, the impact of the RIF had quite the opposite effect, leaving an organization without the skills needed to effectively manage even its severely restricted R&D programs.

The panel also recommended establishing a consolidated technical assessment and analysis capability for DOE, assisted by SERI, to determine the potential of solar technologies, to identify crosscutting research requirements, to evaluate research activities and results, and to provide for industry and university contributions to a coordinated research planning process. It was proposed that SERI be directly involved in the DOE laboratory planning and guidance meetings and that the lines of communication between the senior management of SERI and of the national laboratories involved in solar research be improved. These recommendations had some impact on the program. SERI increased its analysis capabilities, and SERI's senior management became more involved in planning of the other DOE laboratories' renewable energy programs.

At a press conference related to the concerns of the renewable energy advocates, Fitzpatrick sympathized with renewable energy companies' feelings that they needed more support from the federal government but asserted that the short-term outlook was for greater dependence on oil. She felt that many of the problems with renewables could be solved only through long-term solutions. Until the technologies were further advanced, there would be no large market. Creating an artificial demand for solar products would remove the incentive for industry to keep bringing prices down, and the government would have to "keep feeding the baby." With regard to the solar tax credits established a few years earlier, DOE was keeping its options open.

In March 1988 Fitzpatrick noted that "renewable energy technologies currently make important, often under-recognized contributions to the nation's energy supply mix." Renewable energy, mostly hydropower, was contributing about 9 percent of the nation's energy supplies. This contribution was expected to reach about 12 percent of domestic energy production over the next two decades, as industry continued to deploy competitive technology and began to exploit the commercial opportunity of technology emerging from the R&D processes. Promoted to Under Secretary of Energy, Fitzpatrick became the most effective spokesperson for renewable energy to hold that high a post in the administration. Supportive of the federal program within the constraints of the administration's efforts to hold down the budget, she even told congressional authorization and appropriations panels DOE could back higher spending levels than those she proposed if the added money went for what DOE had identified as its core program.

6.2.4 Patent Policy Changes

One of the recurring congressional concerns was that the products of the R&D paid for by the taxpayers rarely found their way into the marketplace. It was frequently stated that the government laboratories held on to new technology that the private sector wanted.

In February 1987 the House Subcommittee on Science Energy Research and Development began formulating legislation to speed the awarding of class patent waivers on technologies developed in DOE laboratories operated by for-profit contractors. PL 96-517, the Bayh-Dole Act, passed in December 1980, established as a general rule for nonprofits and small businesses contracting to do research for the government that, except in cases of national security, patent rights were to stay with the contractor. Subsequently, the president issued an executive order asserting that any rights of the government or obligations of a contract performer be waived or omitted if the agency determined that the public interest was better served. The overall impact of these changes was beneficial; more companies were willing to participate in the solar R&D program.

6.2.5 Energy and National Security

In its report *Energy Security*, released in March 1987, DOE stated that U.S. energy policy should be directed toward increasing domestic oil production by one million barrels per day, providing incentives for energy options that would be the smallest drain on the treasury, and employing energy technologies that would have no serious negative impacts on the economy. Secretary Herrington asserted that while conservation, renewable energy, and nuclear power would play a larger role in the long run, there was no alternative in the short run to increasing oil production. The DOE report projected that the construction of solar buildings was expected to increase steadily, and their energy contribution by 1995 would be "very substantial, particularly from passive systems." By the year 2005, DOE predicted that the use of solar energy in building systems would contribute 0.14 quads of energy from active solar heating and cooling and 2.5 quads from all types of solar systems, including daylighting, although it was vague on just how these contributions would occur, especially in light of DOE's decreasing budget for these programs.

Not surprisingly, the renewable energy industry and lobbying organizations were disturbed by this DOE report, which recommended significant support for the troubled domestic oil industry but not for the equally troubled domestic renewable energy industry. The report made generally favorable predictions about the future role of solar energy, but sought no new initiatives to support those predictions. It noted that early federal programs supported the creation of an industry and market for renewable energy technologies, but held that continued federal support could hinder the long-term development and application of competitive renewable energy technologies. Significantly, DOE did not feel the same way about fossil and nuclear technologies.

In a joint press release in March 1987, a group of renewable energy organizations called the study "unbalanced, short-sighted and patronizing to the consumer's best interests." These groups stressed the need for the United States to develop a balanced national energy policy; they lamented that America was the only industrialized nation without such a policy and called for equity with conventional energy resources in research and development expenditures and tax incentives.

6.2.6 New Legislation

During the Reagan years, the administration generally opposed higher spending in renewable energy and specifically opposed any spending for commercialization efforts. DOE was often challenged for regularly declaring certain renewable energy technologies to be ready for the market (and thus private sector funding) when clearly they were not that far along. Toward the end of Reagan's second administration, draft legislation began to appear that would simultaneously attack the twin problems of relentless budget reductions and the federal government's policy to fund only long-term, high-risk research in renewable energy while funding or even constructing prototypes for other energy technologies. A few examples are discussed below.

To reaffirm the role of renewable energy in U.S. energy policy and to enhance the prospects of significant further commercial application of solar and renewable energy and energy efficiency technologies, Representatives Philip Sharp (D-IN) and Marilyn Lloyd (D-TN) introduced the Renewable Energy and Energy Conservation Commercialization and Development Act (March 1988). In opposing the bill, Fitzpatrick argued that substantially higher levels of program authorizations, by creating false expectations on the part of program participants, could divert attention from promising technologies and from reaching a realistic consensus on technical priorities. Despite strong industry support, the bill died in committee, and the anticipated funding increases never appeared.

Legislation to curb greenhouse gases, in part by boosting spending on renewable energy, was considered to have a good chance of passing Congress because in this area, almost anything done was good for the economy and good for the environment. In August 1988 Senator Tim Wirth (D-CO) sponsored the National Energy Policy Act of 1988, which was intended to deal with climate change associated with the greenhouse effect by aggressively promoting the use of renewable energy resources and conservation and to assist the renewable energy programs by establishing criteria for prioritizing DOE's research and development programs. Although the bill did not pass, it strengthened in Congress the connection between the use of renewable energy technologies and reduced climate change impacts.

Similar comprehensive legislation was introduced in the House by Representative Claudine Schneider (R-RI) in 1988. Among the many topics addressed by this 250-page bill was to establish rigorous least-cost planning principles to be followed by the government. The bill set a three-year authorization for renewables R&D programs at $200 million in FY 1990; $265 million in FY 1991 and $340 million in FY 1992. The bill also required DOE to prepare a report detailing the long-term research, development, and demonstration program and policy options to achieve a doubling, tripling and quadrupling of national renewable energy use by the year 2015. Although it enjoyed bipartisan support, this bill also did not pass. Parts of this impressive bill, however, were taken up in other legislation.

The debate surrounding these bills marked a turning point for the beleaguered renewable energy program and established a context in which the budgets began to grow again.

6.3 The Demise of the Solar Tax Credits

As of 1983, solar tax credits consisted of a 15 percent business energy investment credit on solar hot water heating; a 40 percent residential credit on $10,000 worth of solar purchases and installations, and a 15 percent residential conservation credit on $2,000 worth of purchases and installations. The issue of keeping or eliminating the solar tax credits occupied the attention of the solar industry and solar advocates during the early 1980s. By April 1985, the chances were minimal of passing the Renewable Energy and Conservation Transition Act, designed to extend these credits. The IRS tried to eliminate certain renewable energy technologies from qualifying under this bill by declaring that solar water heating systems that employed a hot water tank with electrical heating backup would not be eligible for the tax credit. The solar industry responded, calling the IRS position absurd, and pointed out that the electric heating element in the tank serves only as backup protection. As the debate on the solar tax credits continued, the industry braced itself for heavy sales as consumers availed themselves of the tax credits before they expired in 1985.

Solar advocates strongly opposed President Reagan's plan to simplify the tax code, which allowed the conservation and renewable energy credits to expire at the end of 1985, while retaining most of the benefits for oil and natural gas well drillers. The plan was presented as "revenue-neutral," and therefore not a way to reduce the federal deficit. The solar industry advocated phasing the tax credits out over several years, to reduce the impact on the solar market. However, other interests tried to restore their tax credits, threatening the revenue neutrality of the plan. At one point, the depletable energy industry was to maintain two-thirds of their enormous tax benefits while the renewable energy industry was to lose most of their very small benefits.

In June 1985 Secretary Herrington, in testimony before the Senate Finance Committee on the president's tax plan, argued that because solar energy systems provided less energy than oil, they deserved less tax relief. Suggesting that solar tax benefits "too often become devices for tax avoidance," he felt these tax subsidies had served their purpose and did not need to be extended. Solar could grow best by relying on the market, consumer behavior, the ongoing federal commitment to conservation and renewable energy, and a simpler, fairer tax system. Herrington supported tax benefits for some portions of the oil, natural gas, and coal industries. He defended the president's plan as "consistent with our national energy policy objectives of assuring an adequate supply of energy at reasonable cost"; the nation's energy policy should be directed toward increasing coal use, developing nuclear power, and encouraging energy efficiency.

It is interesting to note that a 1985 DOE analysis indicated a favorable return on the federal R&D funds spent for renewables. This analysis estimated that from 1975 to 1984 the federal outlays and incentives totaled about $6 billion ($4 billion in R&D and $2 billion in tax credits), while renewable energy systems coming on line during that same period replaced the equivalent of 1.2 billion barrels of oil, worth $39 billion.

Approaching the renewable energy tax credits from another point of view, an OTA study reported that tax credits could make some renewable energy electric power generators attractive to utilities in the 1990s. The study noted that both the renewable energy tax credit and the buyback provisions of the PURPA were crucial in the initial commercial development and deployment of wind and solar power generating technologies.

Although the solar industry knew it had little chance of gaining even a five-year extension of the renewable energy tax credits, it also knew that an abrupt elimination of the residential tax credits—the equivalent of a 40 percent price increase overnight—would have a devastating impact. Furthermore, it was predicted that unless the federal renewable energy tax credits were extended, states could not be expected to renew their own tax credit programs, leading to a decline in the nation's capability to cope with future energy shortages.

By November 1985, the only chance for an extension on the renewable energy tax credits was as a provision in a possible "tax credit clean-up bill," which would allow a host of benefits to remain in place long enough for Congress to include them in a comprehensive tax reform act in 1986. After significant maneuvering, the House and Senate seemed ready to support a 30 percent tax credit in 1986, and 20 percent in 1987 and 1988 for residential solar water heating, solar space heating, and PV equipment, as opposed to the existing 40 percent credit. But time ran out and the tax credits died.

Some solar companies began diversifying while others left the field, creating what would become to be known as "solar orphan systems." SEIA reported that the distribution network for solar energy systems was beginning to crumble. Some solar companies, however, remained optimistic about their prospects for staying in business even after the credits disappeared, ironically because they felt that so few consumers ever knew about the tax credits in the first place.

In early 1986 the Senate Finance Committee staff released a summary of a tax reform bill containing a measure to restore the solar benefits at reduced rates, and the hopes of the solar industry began to rise. In March of 1986 three plans emerged with respect to renewable energy tax credits: (1) no credits—the original Administration position; (2) modest and time-limited credits—the House plan; and (3) extensive, reduced-value credits —the Senate plan. The House bill did not extend the credits for wind energy or ocean thermal systems, while the Senate bill did not extend the residential credits. However, the Senate was increasingly reluctant to change the tax code, and the House was concerned about the loss to the treasury. Under the compromise tax reform bill, businesses investing in solar thermal, ocean thermal, geothermal, and biomass energy equipment would be eligible for retroactive renewable energy tax credits. The 40 percent residential solar energy credit and the 15 percent wind energy credit were not extended. The business solar energy investments approved for extension were 15 percent in 1986, 12 percent in 1987, and 10 percent in 1988. The R&D credit would be maintained another three years, but the rate would be cut from 25 percent to 20 percent; the standard investment credit was repealed. The compromise tax reform bill passed the House and the Senate and was signed by President Reagan.

As things settled down, it became clear that the solar industry would have to undergo a major transformation. Purchaser confidence in solar water heating was badly shaken by reports of fraud against homeowners and the IRS. The industry would have to first shrink to align itself with a smaller market, find ways to reduce the price of their products, primarily solar water heaters, and then develop new marketing strategies.

6.4 Overview of Renewable Energy Budgets during the Reagan Administrations

Perhaps the most effective way to summarize the impact of the two Reagan administrations on the renewable energy programs is to briefly review the annual budget ritual that went on between Congress and the executive branch. When all of the FY 1981 budget battles were over, renewable energy received an appropriation of $549 million, almost a 200 percent increase above the administration's original request. The reader may remember that this was the era when all Reagan budget submissions were declared "dead on arrival" by the leadership of the Democrat-controlled House. Energy budgets, especially, were viewed in this manner and renewable energy programs benefited from this congressional position and fell into a yearly pattern of low request followed by a substantial increase added by the Congress.

The FY 1982 renewable energy budget was finally appropriated at $259 million. This amount was later reduced by 12 percent, to $221 million, as the result of an across-the-board reduction in all federal programs requested by the president and agreed to by Congress.

In its FY 1983 budget request of $73 million for renewable energy, a reduction of $148 million from the previous year, the administration totally eliminated funding for the active solar heating and cooling programs. This action served as the catalyst for undertaking this series of books documenting the results of the federal programs in solar heating and cooling. The fear was that the programs and the people with the knowledge of the program accomplishments could disappear overnight. After the usual budget hearings and debates, the funding for renewable energy was restored to $197 million, with $12.5 designated for the solar heating and cooling programs.

For FY 1984 the administration asked for $87 million for renewable energy, a surprising increase of $14 million above the previous year's original request, although once again, the active solar heating and cooling and passive solar programs were zeroed out. In justifying this move, DOE argued that the technologies were well enough developed to allow industry to commercialize them as they saw fit. Congress did not agree with this argument and restored the renewable energy budget to $177 million, which included $16.5 for the solar heating and cooling programs.

FY 1985 saw the same dance as in previous years, with one difference, however. Before leaving DOE for his new post of Secretary of the Interior, Hodel broke with the administration and restored most of the funds that OMB tried to remove for renewable energy. As a result, renewable energy was finally appropriated $178 million in FY 1985, essentially the same amount as the previous year.

Although budgets continuously declined during these years, FY 1986 witnessed a change in the usual give-and-take between Congress and the administration. This time, when the administration again proposed a cut in the renewable energy budget to $143 million, Congress agreed to its request. Whether this capitulation represented exhaustion or the realities of ever-growing budget deficits, Congress finally saw eye to eye with the administration on a budget for renewable energy. In the short space of six years, solar (renewable) energy budgets had plummeted from $549 to $143 million. In FY 1986, the solar buildings program got $1 million less than the requested $9.5 million, but Congress transferred $1 million from the DOE conservation sector's buildings and community systems budget to the solar buildings budget to address the special problem of cooling in hot humid climates for single and multifamily buildings.

The original DOE request for the renewable energy program for FY 1987 was $199 million. OMB's reaction to that request was to reduce it by almost 60 percent, explaining that at the reduced budget level there would be adequate funds for a continued, limited federal role in support of long-term, high-risk R&D. Almost every line item in the renewable energy budget was cut, some by more than 50 percent. The DOE request of $4.7 million in FY 1987 for solar building energy systems for materials and components research and systems R&D was $3.5 million below the previous year. Once again, DOE justified this budget by stating that the "federal funding is concentrated on basic and applied research and development which is conducive to informed decision making by the private sector regarding the development of a reasonably broad range of potentially competitive technologies." DOE's mission was to concentrate on scientific and engineering research to better understand basic resource and technology phenomena, on methods to employ these resources for producing competitive and useful forms of energy, and on important crosscutting technologies for storage of energy and for reliable and efficient management of the energy supply network.

Congressional reaction to the drastically reduced FY 1987 renewable and conservation programs budget was, as usual, negative. House Energy and Commerce Committee Chairman John Dingell (D-MI) commented that the proposed DOE budget would threaten the nation's energy security and that it did not reflect the balanced energy budget that Congress had advocated. The House Energy Development and Applications Subcommittee voted to increase spending on renewable energy from the administration's request of $72 million to $122 million.

In September 1986, under threat of a presidential veto, the House approved by a single vote a $560 billion omnibus spending bill for FY 1987 that included funding for DOE. Meanwhile, the Senate Appropriations Committee voted for a continuing resolution bill at $556 billion, with $126 million for the renewable energy programs. By the end of October, 1986, with a few modifications, the president signed an appropriations bill that included $122.5 million for DOE's renewable energy program, with $6.9 million earmarked for the solar buildings program.

By the summer of 1987, the decline in federal funding of DOE's civilian R&D programs, particularly for renewable energy, had been devastating. The intellectual resource base for solar energy was being dismantled and dispersed. Freshman Senator Wyche Fowler (D-GA) was displeased with DOE over its policy of funding only long-term, high-risk R&D, and even then only at minimal levels. Senator Fowler held hearings in which he sought to outline a federal renewable energy policy that could make more effective contributions to the development of renewable energy technologies, taking into account the realities of tight federal budgets and strong international competition. These and similar hearings established a premise upon which the budgets finally began to increase during the Bush administration.

Despite Senator Fowler's concerns, and to no one's surprise, DOE's proposed FY 1988 renewable energy budget request was once again reduced. The request was $71 million and included cuts in all renewable energy programs; funding for R&D programs would continue to decrease across the board, with emphasis on long-term generic R&D rather than near-term product/technology-specific R&D. Regarding solar buildings, DOE stated that by turning the component phase over to industry (which had already been done years ago), it could concentrate the research effort on materials and advanced glazings for passive systems, thereby saving federal funds.

Senators Mark Hatfield (R-OR) and Dan Evans (R-WA) told DOE that its budget did not put enough emphasis on the role of renewable energy in ensuring the nation's access to secure energy resources or in protecting jobs in the nation's energy industries. DOE responded that it recognized that renewables over the long term would play a major role, but the budget focused on what could be done year by year.

During the same period, in its budget presentation to the House Science, Space, and Technology Subcommittee on Energy Research and Development, DOE stated that its goal was to increase the energy contribution of solar technologies "to the point where 70–80 percent of space heating, hot water, cooling and lighting requirements of many different types of buildings could be supplied at costs that were competitive with conventional fuels." Somehow, DOE seemed to feel this goal could be achieved with the small R&D support budgets they had presented.

In developing the authorization for the FY 1988 renewable energy budget, legislation to help the United States retain its technological lead in solar energy was introduced by Senators Dale Bumpers (D-AR), Wyche Fowler (D-GA), Dennis DeConcini (D-AZ), Spark Matsunaga (D-HI), and Alan Cranston (D-CA). By this point, control of the Senate had reverted back to the Democrats, and they hoped that, with House support, their positions would prevail. The bill authorized funding at current levels for DOE's ongoing renewable energy programs for the following three years. The proposed legislation would also direct the existing federal loan programs for small business and for exports to be more responsive to the needs of the solar and renewable energy industries. Senator Fowler proposed to set up multiyear authorizations for both renewable energy and energy conservation; he also wanted DOE to set up six joint ventures between government and industry working on demonstration projects and marketing programs. The administration, not surprisingly, was absolutely opposed to federal financing of solar demonstrations and marketing and, ultimately, neither of these bills passed.

In the spring of 1987, during House Science Energy Research and Development Subcommittee hearings on the recent accomplishments, current status, and potential of renewable energy development in the United States, many witnesses said that if the renewable energy budget were reduced any more, there would be little point in maintaining a national program. Noting that the United States had spent approximately $4.5 billion on R&D in renewable energy, the subcommittee said it needed a better understanding of what this money had bought in increased resource utilization. There were questions as to which of the renewable energy technologies should be saved if only a few were spared. All witnesses asked this question declined to give an answer, except Donna Fitzpatrick, the Assistant Secretary for Conservation and Renewable Energy. Her reply was, "You are asking me which of my children will grow up to be the most valuable? First tell me what needs you have for this resource—heating, cooling, lighting, fuel, or electricity—and then I will tell you which renewable energy technology will best meet those needs." The discussion ended there. The FY 1988 solar budget was finally approved at $97 million, and the conferees agreed to reauthorize the Solar Bank.

As planning for the FY 1989 renewables budget began, everyone involved in the battles and down-to-the-wire deals on the FY 1988 renewables budget braced for another round of proposed cuts. The administration cited the still-growing federal deficit and concern that the reconciliation bill would not reduce federal spending by as much as was anticipated. It was clear to solar advocates that while the Congress managed to increase the renewable energy budgets each year above the level requested by the administration, the size of those increases was getting smaller and smaller. The contradiction of DOE's budgets during the Reagan administration is that they asked for spending cuts while praising the promise of renewable energy. For FY 1989, despite the claim that it was offering a significant, stable commitment to renewable energy, DOE's budget request was just over $80 million, reduced from $97 million in FY 1988. DOE argued that the proposed budget would allow the department to "advance the engineering and scientific understanding of solar energy" and to establish the technology base from which private industry could develop economically competitive solar-related products and designs for use in existing and new buildings. It again proposed to do this by reducing funding for the solar buildings program.

Concerned about the annual budget deficit, and unwilling to battle with the White House over this relatively unimportant issue, Congress settled on a FY 1989 solar budget of $92 million. In appropriating the FY 1989 solar budget, Congress noted that most solar and renewable technologies had not been developed to the point where the private sector alone could be expected to carry forward with their development; it again asked DOE to report, within six months, on its activities and plans to bring DOE research results to the marketplace. This request was a typical tactic utilized by Congress during this period to underline its displeasure with progress in the solar program.

In legislation approved by the House Science, Space, and Technology Subcommittee on Energy Research and Development in September 1988, authorizations for DOE's solar energy budget would rise in FY 1990, FY 1991, and FY 1992. But the Senate Energy Committee, and the House and Senate Appropriations Committees did not go along with this. Although the desire to change DOE's funding pattern remained alive on

Capitol Hill as the end of the second Reagan administration approached, in the end-of-the-year rush to adjourn, the legislation providing the next three years of authorizations for solar energy at substantially higher levels than those for the current fiscal year failed to clear the 100th Congress.

6.5 The Slow Recovery Years

Because this volume was originally expected to be completed and published before the end of the second Reagan administration, no author was assigned the task of carrying the history beyond 1988. Each year of delay in the series' completion, however, has made this shortcoming more obvious. While the current authors have remained peripherally involved in the solar heat technologies program in the 1990s, we are not in a position to write an insider account of its evolution. We will, therefore, confine ourselves to reporting some of the most apparent trends and events that have taken place since the end of the Reagan presidency.

6.5.1 The Shift from Energy Security to Environmental Concerns

The driving force for the solar energy program that evolved in the 1970s was national concern over limitations of, and secure access to, the primary sources of fossil fuels. By the end of Reagan's second term, the finiteness of fossil fuels was all but forgotten, and U.S. relations with Saudi Arabia, holder of the world's largest known oil reserves, were good. The price of gasoline was low and winters were mild; barely anyone in the United States was concerned about reducing our dependence on fossil fuels. These concerns, however, had been replaced to a large degree by others, about the immediate and long-term impact of the increasing use of fossil fuels on the local and global environment. Many energy policy makers expected these widespread environmental concerns, in concert with the political repercussions of the Persian Gulf War, to trigger a serious commitment to the transition from a fossil-fuel-based economy to a sustainable energy economy, based to a large degree on renewable energy resources such as solar energy and inexhaustible resources such as geothermal heat and nuclear fusion. U.S. government policy, however, has yet to come to grips with either the environmental threats or the certainty of the diminishing availability of oil and natural gas, and thus the certainty of rising costs for conventional energy.

6.5.2 The Impacts of the Bush Administration

George Bush came to office in 1989 with the perception of being more forward thinking on energy and environmental policy than his predecessor. His Chief of Staff, John Sununu, however, seemed concerned primarily with politics and fund-raising; he found the fund-raising ground far more fertile in the large, established energy production and industrial sectors than in the grassroots energy and environmental activist communities.

Bush's Secretary of Energy, Admiral James D. Watkins, a nuclear engineer from Admiral Hyman Rickover's nuclear submarine navy, had a reputation for strong leadership and management skills. Although developing a new energy policy was Watkins's avowed first goal, he took charge of DOE at a time when its weapons program, which then constituted 53 percent of its budget, was in disarray, beset by safety, contamination, and disposal problems so severe that it could no longer produce nuclear weapons within legal safety regulations. Watkins's tenure was thus preoccupied with correcting the operations and management problems within the defense side of the department.

When Watkins eventually produced a new *National Energy Strategy* report, it was generally disappointing to the renewable energy and environmental communities, who regarded it as a strategy for increasing and prolonging production of energy from fossil fuel and nuclear resources rather than for moving toward a sustainable energy future. The energy bill that eventually passed Congress, PL 102-486 the Energy Policy Act of 1992, was even weaker in its support of renewables and energy efficiency than Watkins's "strategy" report.

Renewable energy technologies and energy efficiency research and development did, however, fare better under Watkins than under his predecessors in the Reagan administration. Watkins's choice for Assistant Secretary for Conservation and Renewable Energy, J. Michael Davis, was a former program manager and branch chief in the DOE solar heat technologies program, a division director at SERI, and the owner of a solar and energy efficiency equipment installation company in Colorado. His appointment was applauded by the solar community, who looked to him to rebuild the renewables program and increase its budget.

Davis and Watkins reversed the long, slow decline of government support for renewable energy technologies; between 1989 and 1993, the renewable energy budget increased by a little more than 80 percent. As shown in figures 1.3 and 1.4, however, the Bush administration budgets were still only a fraction of the Carter budgets. The greatest beneficiaries of the budget increases were the photovoltaics, wind, and biomass programs. The solar buildings program continued to decline and the solar thermal power program did not fare much better.

Like so many of his predecessors, Davis reorganized the Conservation and Renewable Energy Office, this tme along market sector rather than technology lines (see figure 1.8). In addition to the Office of Utility Technologies and the Office of Building Technologies, which contained all the former solar heat technologies programs, there were offices for industrial energy efficiency, transportation, and state and local programs.

Dividing the solar programs between the utilities and buildings offices had especially unfortunate consequences for the solar buildings program. The program was split into two parts, active solar and passive solar. Each was small, under funded and unappreciated in their new departmental organizations, where all the upper managers had come from the energy conservation side of the house. It was especially ironic that the solar heating and cooling programs were placed in the end-user/builder-oriented side of the office at just the time when these programs were realizing that the best realistic market delivery mechanism for these technologies involved the utilities. Had it been possible to include the active solar program in the Office of Utility Technologies instead, it certainly would have fared better organizationally and would likely have done better in budget allocations as well.

Under Davis, "commercialization" was restored to the DOE dictionary, although he often preferred "market development" or "market conditioning." Some very successful industry-government joint ventures were finally launched in the photovoltaics, solar thermal, and biomass programs; integrated resource planning (IRP) for utilities became a department mantra; solar and wind energy tax credits were made permanent and performance-based; and SERI, renamed the National Renewable Energy Laboratory (NREL) fared very well.

6.5.3 The Impacts of the Clinton Administration

The election of the Clinton-Gore ticket in 1992 once again raised the hopes of renewable energy and environmental advocates to new highs. Gore was a staunch advocate of a sustainable, renewable energy economy, and with the publication of his best-selling book *Earth in the Balance*, he had become the leading spokesman for a national environmental policy as well. Although energy and environment had not been major issues of the 1992 campaign, the President-elect did promise that the DOE would play a major role in stimulating the economy and creating jobs. His advisors promised an economic and energy policy to meet three prime national goals: economic viability, environmental quality, and strategic security.

As part of his economic plan, Clinton considered a carbon tax, as advocated by Gore and Tim Wirth, but eventually rejected it as "too hard on coal states." He proposed instead, a Btu tax based on the energy value of fuels used to produce heat and power. The tax of 25.7 cents per million Btu would apply to energy from all sources other than solar, wind, and geothermal resources. Gasoline and other refined petroleum products would be taxed at a rate of 34.2 cents per million Btu. The proposals were immediately attacked by special interests and ridiculed by congressmen who did not believe the American people knew what a Btu was. (They were certainly correct in that point.) Although renewable energy advocates and environmentalists praised the Btu tax as a step in the right direction, few believed that the tax was large enough to significantly influence energy use patterns. After a series of compromises that effectively gutted the Btu tax proposal, the administration settled for a 4.3 cents per gallon increase in gasoline tax. Hope for a strong energy and environmental administration faded quickly.

President Clinton's expected choice for Secretary of Energy was Tim Wirth, former Senator from Colorado and leading advocate of the need for global warming legislation. However, the president finally appointed Hazel O'Leary who had more experience in the energy field than most of the other previous Energy Secretaries. O'Leary was considered a good choice by most factions, with the possible exception of the defense community, who questioned her understanding of the department's weapons work.

O'Leary's DOE has taken some steps toward more rational energy policies including a new strategic plan that emphasizes industrial competitiveness, energy resources, national security, environmental quality, and science and technology. The department is also a major player in Vice President Gore's Climate Change Action Plan, and has induced many utilities to participate in a voluntary "Climate Challenge" to reduce greenhouse gas emissions to 1990 levels (or below) by the year 2000.

Under O'Leary, budgets for renewable and energy efficiency programs have increased, but only at about the same rate as they did during the Bush administration; budgets for the solar heat technologies have actually dropped slightly. O'Leary's choice for Assistant Secretary for Energy Efficiency and Renewable Energy, Christine Ervin, a former director of the Oregon Energy Department, was a late appointee at DOE and has yet to have any noticeable impact on the solar heat technologies programs or activities.

7 Significant Results of the Federal Solar R&D Programs

Charles A. Bankston

This chapter offers personal reflections on the content of this series of volumes and, in some cases, on events that have taken place or information that has become known since the volumes were written. This is not a research paper and does not provide references to support its observations. It is, however, the best effort of one who was not only involved in the planning, writing, and editing of the series from its inception, but who was also involved in much of the research upon which the series is based and is still an active solar energy researcher and consultant.

I will stress the accomplishments of the government solar programs from the mid-1970s to the mid-1980s ("the solar era"), drawing primarily from the material in the other volumes but also including material that is more recent or, for some other reason, was not included in the series, and occasionally calling attention to things that were not achieved or were only partially successful. Many things were not accomplished simply because the solar heat technologies program changed direction or ran out of time and/or money. In a few instances, ideas, technologies, projects, or programs were not successful because they were incorrect, inappropriate, or ill conceived. These are also important results, and the authors and editors of this series have reported these valuable lessons along with the positive results.

The chapter is organized into sections corresponding to the nine volumes of this series that deal primarily with the R&D programs. The first two sections cover the general studies of solar resources and solar economics that were reported in volumes 2 and 3, respectively. The third section deals with solar technologies or components that were reported in volume 5. Work related to solar buildings, reported in volumes 4 and 6–9, is covered next. Work on solar thermal power and industry, which was to have been covered in the series but was dropped for reasons of time and cost, is discussed briefly in the next-to-last section. The final section deals with the commercialization activities that are covered in volume 10. Although the sections correspond to other volumes, there is no attempt to use the same titles or organization of the actual volumes.

7.1 Solar Resource Analysis

Everyone with even a casual interest in solar energy is aware that in order to design, plan, or evaluate solar technologies, projects, or programs, one must have a relatively precise quantitative knowledge of the solar radiation that reaches the earth in locations of interest. A solar radiation database is thus a prerequisite to any effective attempt to utilize the sun as a major source of energy. But when the solar era began, the knowledge base of solar radiation in the United States was woefully inadequate, and databases for many parts of the world simply did not exist.

This vital activity was authorized in 1974 as part of PL 94-473, the Solar Energy Act, which directed "a solar energy resource determination and assessment program with the objective of making regional and national appraisals of all solar energy resources including data on insolation, wind, seas' thermal gradients, and potentials for photosynthetic conversion." Prior to the Solar Energy Act, the only systematic collection of solar insolation data was undertaken by the National Weather Service (NWS), which regarded solar radiation as just another manifestation of weather. Outside the solar research community, there was little appreciation for the value of the natural resource represented by the solar energy that could be collected on the earth every day.

The responsibility for implementing the resource assessment program initially fell on NSF, which quickly recruited the National Oceanic and Atmospheric Administration (NOAA) and NWS to make the actual measurements. Over the years, program responsibilities were shifted to ERDA and subsequently to DOE. The effort nearly vanished in 1981, with the budget reductions of the Reagan administration, but survived and has continued at a modest level. Although the administrative responsibility shifted, much of the technical expertise and research eventually came to reside at SERI (now NREL). Roland L. Hulstrom, editor of *Solar Resources* (volume 2 in this series), directed that effort for a good part of the solar era.

The reader unfamiliar with the status of the solar resource assessment may be surprised by the sparsity and inaccuracy of data. Even today, coverage is quite limited, many of the most important data sets are obtained from empirical models rather than measurements, and the accuracy of the primary measurements is rather poor. Nevertheless, the situation is vastly improved over that which existed prior to 1974. A solar researcher or designer today can find measured historical data that covers a period of thirty years for 26 measurement sites (SOLMET stations), and synthetic, or modeled, data for 222 (ERSATZ station) sites. There is also an artificially created, typical meteorological year (TMY) data set that is available for some 235 sites to predict the behavior of a system in future years. Current field instrument accuracy is still relatively poor—no better than ±5 percent, and the accuracy of the models used to fill in missing results and to transform one solar radiation measure to another is considerably worse. Still, despite the annual statistical variations due to normal weather uncertainties and unexpected geophysical events (such as volcanic eruptions), the meso- and microclimate variations, and the uncertainty in the response of systems to the radiation, the limitations of the resource data are not a major problem for the solar energy system designer. However, the researcher and the analyst who need more accurate and detailed data for their particular site(s) must include solar radiation measurement in their research programs.

Since *Solar Resources* was completed, NREL has released the *National Solar Radiation Data Base for the United States*, which provides historical data for the period from 1961 to 1990 and contains data for 250 sites (see figure 7.1). New TMY data is promised by NREL but has not been released.

7.2 Solar Energy Economics

The close ties between the cost and availability of energy and the world and national economy's well-being are no longer a subject of debate. The events and recessions of the 1970s and 1980s provided convincing evidence of the world's reliance on cheap and readily available sources of energy. Solar energy is free and abundant in virtually every part of the world. Unfortunately, the means of using it constructively to produce goods and services are not free. Solar energy economics are characterized by high capital costs and low operating costs. This is very different from the economics of fossil fuel energy resources, such as oil, natural gas, and coal, where the cost of extracting and delivering the resource may be very small compared to its market value. Which means that indigenous, renewable energy use will influence national and global economies in profoundly different ways than conventional energy sources. The full ramifications of these differences are still unknown and the subject of much conjecture and considerable disagreement.

The much narrower question of how to evaluate the economic feasibility of a solar energy project in relation to other, conventional, energy

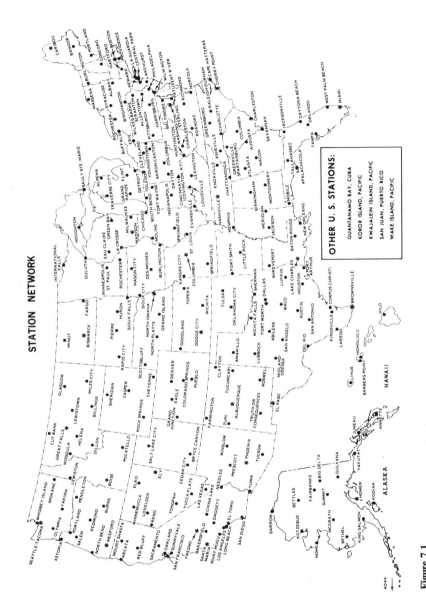

Figure 7.1
Locations for which SOLMET and ERSATZ data are available.

options has, however, been given considerable attention in the government solar heat technologies program as well as in the private sector. This is basically the question that is addressed in *Economic Analysis of Solar Thermal Energy Systems* (volume 3 in this series), edited by Ronald E. West and Frank Kreith. To a lesser degree, the volume also discusses some of the work done early in the program on the larger issues of market potential and the effect of government policy on the market.

Except for some examples, the authors of volume 3 do not attempt to determine the economic feasibility of specific projects or even specific technologies. This is fortunate because the book was the first completed in the present series, and it contains only the results and work available up to 1985. Much has changed since 1985. Market projections, cost goals, and so on made early in the program and reflected in volume 3 have been rendered useless by the decline in oil prices (in terms of constant dollars), the glut of natural gas, and changes in government policy on energy. The tax credits and exemptions that were available for many types of solar projects have been eliminated or changed, the price of oil and natural gas has fallen and is now at or near twenty-year lows in real terms, and the costs of some solar technologies have fallen dramatically, while others have increased or remained level in current dollars.

Nevertheless, even though the input data have changed, the basic methodology described in volume 3 has become better established and more widely used. Utility companies now are often required to include renewable and energy efficiency measures on both sides of the meter in their analysis and optimization of new capacity measures (integrated resource plans). Some of the societal costs that were only talked about at the beginning of the period are now tentatively quantified, and in some states, set by rule.

The analysis of cost trends over the period studied in volume 3 was inconclusive. Later data have not been thoroughly analyzed, to this author's knowledge, but a reduction in real cost of most solar thermal technologies is clear. Another important question left open in volume 3 because the analysis was, and still is, incomplete, is the impact and effectiveness of government incentives. The residential and commercial tax credits that were widely used to promote solar equipment sales were always controversial. Although there is still no definitive evaluation of their economic and energy impacts, the consensus is now that they did more harm than good.

The evaluation and allocation of risk and the benefits of flexibility in the analysis of energy projects have become important topics in recent years. One of the major risk factors generally not included in economic analysis is the risk of fuel price volatility. Methods to account for this risk, and some of the advantages that solar technology offers in modularity and flexibility, have been developed recently and could have important implications because the marketing of all solar technologies will probably move toward the utility sector.

7.3 Solar Heat Technology

Although NSF, AEC, and NASA had all supported solar technology R&D as well as solar energy projects, ERDA was the first to systematically analyze the needs of the solar technologies for buildings and industrial applications, and develop a focused research and development program to fulfill these needs. In the fall of 1975 and 1976, ERDA sponsored a series of technology assessment meetings that included representatives of industry, universities, and government. This assessment was the basis for the preparation of the first comprehensive R&D plan, issued in 1976 as ERDA 76-144, *National Program Plan for Research and Development in Solar Heating and Cooling*. This plan identified ten paths that could logically be followed to achieve the three major building applications objectives: service hot water, space heating, and space cooling; and eleven paths for the four agricultural and industrial application objectives: shelter heating, drying, hot water, steam (100 to 200°C), and cooling. The paths included all of the tasks that would be necessary to achieve the goals, starting with the identification of the resource, its capture, storage, energy conditioning, optimization, and control. The buildings applications paths included more than 300 individual tasks. Tables 7.1 and 7.2 show the buildings applications paths and task structures respectively.

These tasks were implemented through a series of solicitations issued in 1977. The nine applications solicitations (RFPs and PRDAs) for the buildings sector resulted in nearly 2,000 proposals and more than 200 contracts with individuals, business, and universities to supplement the R&D programs conducted within the government laboratories that originated from unsolicited proposals funded by all of the agencies. This outpouring of research ideas and the funds to pursue them kept the

Table 7.1
Path descriptions used in the *National Solar Heating and Cooling R&D Plan*

Application	Designation	Path description
Service Hot Water	W1	Liquid-heating collectors
	W2	Air-heatng collectors
Space Heating	H1	Solar-assisted heat pump
	H2	Direct solar heating of space or structure
	H3	Air-heating collectors
	H4	Liquid-heating collectors
Space Cooling	C1	Concentrating collectors with absorption or Rankine cycle chiller
	C2	Advanced nonconcentrating collectors with absorption or Rankine cycle chiller
	C3	Flat plate collectors with desiccant chiller
	C4	Evaporative and night-effect cooling

solar research and development community busy for the next four years and resulted in many of the accomplishments described in this series. Figures 7.2 and 7.3 are time lines of some of the accomplishments of the Solar Heating and Cooling Program in active and passive technologies respectively.

Virtually all the accomplishments described in volume 5, *Solar Collectors, Energy Storage, and Materials*, resulted from the R&D Plan. This massive volume was edited by Francis de Winter, a former student of Hoyt Hottel at MIT and a veteran of spacecraft power research as well as solar thermal technologies. Mr. de Winter is well known in the solar community for his research and his involvement in national and international solar society affairs.

7.3.1 Materials

Much of the history of technological advances is stated in terms of discovery of the materials that made them possible. From the Bronze Age to the Silicon Age, our times have been shaped by the materials that were available to fabricate tools and machines. The advances in solar thermal technologies are a result of the ingenuity of not only inventors and designers but also the developers of the materials that made solar products better, more efficient, and cheaper. Often advances in one area are made possible by materials developed for an entirely different purpose. For example, electrodeposited black chrome, which has highly selective

Table 7.2
Task classification for the *National Solar Heating and Cooling R&D Plan*

Major classification	Subclassification

I. Solar collectors	A. System studies related to collectors
	B. Theory and basic phenomena
	C. Solar liquid-heating collectors
	D. Solar air-heating collectors
	E. Advanced nonconcentrating collectors
	   a) Evacuated tube collectors
	   b) Heat pipes
	   c) Honeycombs
	   d) Others
	F. Concentrators
	G. Reflectors for flat plate collector augmentation
	H. Materials
	   a) Selective surfaces
	   b) Glazing
	   c) Glazing surface coatings
	   d) Sealants
	   e) Coolants
	   f) Insulation
	   g) Nonmetallic films
	J. Tests and standards

II. Thermal energy storage and heat transfer	A. System studies related to storage B. Piping, ducting, heat exchangers C. Water tank storage D. Rock bed storage E. Phase change storage F. Chemical storage G. Tests and standards
III. Solar air-conditioning and heat pumps	A. System studies related to air-conditioning, heat pumps B. Absorption chillers C. Rankine cycle vapor compression chillers D. Evaporative and night-effect chillers E. Desiccant chillers F. Other chillers G. Heat pumps
IV. Systems and controls	A. Integrated active solar energy systems B. System design methods C. Controllers, valves, and actuator components D. Instrumentation and data acquisition E. Passive and hybrid solar energy systems F. System studies
V. Non-engineering aspects of solar heating and cooling	A. Economic and financial analysis B. Consumer attitudes and behavior C. Marketing, architecture, and construction D. Energy utilities E. Legal, regulatory, and legislative considerations F. Education and training G. Social and environmental considerations

Significant Results of the Federal Solar R&D Programs

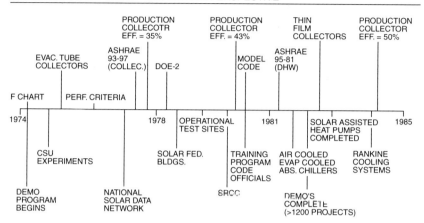

Figure 7.2
Historical perspective of accomplishments of the National Solar Heating and Cooling R&D Program for active solar applications.

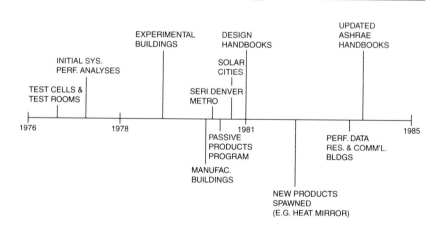

Figure 7.3
Historical perspective of accomplishments of the National Solar Heating and Cooling R&D Program for passive and hybrid solar applications.

radiation absorption properties that are attractive for solar collector absorber surfaces, was developed as an aesthetic finish for metal furniture. But more often, the materials that make a difference are sought out and improved for specific products and applications.

Solar thermal technology, which at one level can be implemented with the simplest of building materials, also looks to materials science for the high-tech materials that can make more sophisticated applications possible and cost-effective. Although the directed materials research of the late 1970s and early 1980s resulted in laboratory advances that were substantial, unfortunately not many of those advances found their way into commercial products because of the loss of solar markets and the retrenching of the solar industry. This makes the documentation of the solar materials R&D doubly important.

Perhaps the most important advances in materials for solar applications came in the development of materials with special optical properties that enhance the optical performance and overall efficiency of solar thermal devices. Whereas solar collectors and water heaters of the early twentieth century relied upon flat black absorbers (paints or soot coatings that emit as strongly as they absorb), the designers of solar collectors today can choose from a variety of highly selective coatings that absorb virtually the entire solar spectrum, but emit weakly in infrared (IR). Selective surfaces may now be produced by a variety of means including electrodeposition, vapor-deposition, sputtering, chemical coating conversion, and composite paints. The designer can choose from simple paints for low-temperature applications to ultrasophisticated vapor-deposited or sputtered coatings that are efficient and stable at temperatures approaching 930°F (500°C). The transmissive materials available for collector and building glazings also have experienced a revolution. Although the ideal glazing—one that would combine low weight and low cost with ideal optical properties—still eludes us, there have been some important and even spectacular developments in optical properties. These range from reducing the amount of iron in glass to improve its transmissivity to developing photochromic and electrochromic coatings that allow the window's or glazing's transmission to be controlled by the incident sunlight or by an external control device. The optical properties of glazing have also been enhanced by the development of processes to reduce the reflection losses (antireflective coating and finishes) or the IR losses (heat mirrors). There have also been important strides in the performance, durability, and cost effectiveness of reflective surfaces for concentrating and nonconcentrating collectors. Both silver and aluminum on polymer film may be applied to either glass or metallic substrates. Back- and front-surfaced thin glass mirrors may also be bonded to structural shapes. Advances in protective coatings have made front- and back-surfaced mirrors more rugged and longer-lived. A big advance is the development of stressed (or stretched) membrane steel or polymer mirrors for concentrating collectors and heliostats. These advances are expected to have a major impact on the cost of mirrors produced in high volume.

A great deal of research was devoted to finding or developing suitable plastics to reduce the weight and cost of solar collectors of all types. Although this work has been relatively successful for low-temperature collectors used in swimming pool heating or agricultural applications, it has not resulted in a marketed, long-lived, medium-temperature collector for water or space heating application.

7.3.2 Collectors

Collectors are the heart of all solar heating systems. Solar heating applications cannot hope to be successful until the collectors are capable of delivering thermal energy at a cost that is competitive with conventional energy resources. Therefore, the development of cost-effective solar collectors is the key to all applications.

As with materials, the R&D advances made during the government programs' peak years were greater than the advances in marketable products. While the advances have not stopped, they have slowed considerably, especially in the building applications field, where government participation has practically vanished. Support of R&D for solar thermal power and industrial applications has been more persistent, and so have technological advances for the collectors used in those fields.

7.3.2.1 Solar Collector Concepts
More than a thousand solicited and unsolicited proposals, most of them involving innovation or incremental advances in the collector technology, were reviewed and evaluated over the course of the solar heat technologies program. This led some of the reviewers to observe that there is truly "nothing new under the sun." Inventors, researchers, or investors should be aware that there is very little in solar technology that has not been thought of, and quite likely evaluated. This is not to discourage new thinking, but rather to encourage

those who would advance the state of the art to be sure they are aware of what has been done. The material in this series is undoubtedly the most complete guide to the vast literature on solar heating systems. (Chapters 2, 3, and 9 of volume 5 and chapter 17 of volume 6 in this series are especially good starting points for readers interested in new concepts.)

7.3.2.2 Optical and Thermal Analysis The ability of engineers and scientists to analyze, predict, and optimize the performance of solar energy conversion systems is critically important to the development of the technology. Although much was already known about the performance of simple flat plate and some concentrating collectors, the advances during the solar era have made it possible to accurately predict the performance of nearly every conceivable solar collector configuration. Advances in computers have even made detailed analysis of complex optical and thermal system practical. Still, many practitioners rely on simplified, linear methods developed to accommodate the limitations of the calculators and slide rules of the 1950s. The models developed by Hoyt C. Hottel and coworkers at MIT in the 1950s for flat plate collectors are still in widespread use. In fact, they are usually further simplified by subsuming many of the details into experimentally derived constants. The wisdom of this practice may be somewhat suspect because even in the 1950s, according to Hottel, it was possible to predict collector performance within about 2 percent accuracy using hand calculation methods, although it was necessary to include geometric details (such as edge-shading conduction losses) and to include some second-order effects. Not many test laboratories have been able to measure the efficiency of solar collectors to this degree of accuracy.

Advances in the optical theory and analysis of concentrating solar collectors have had a substantial impact. Computer ray trace techniques have led to the development of a new field of optics—nonimaging optics—and completely new optical devices, such as the compound parabolic concentrator (CPC) collector. The CPC collector can be used as a fixed, low-concentration primary collector with a wide acceptance angle, or as a secondary collector to achieve high overall concentration ratios in combination with less accurate (and less expensive) primary concentrators. CPC devices have been used to achieve incredible concentration ratios (70,000) and absorber temperatures higher than the temperature of the surface of the sun. All these developments are based on the work of

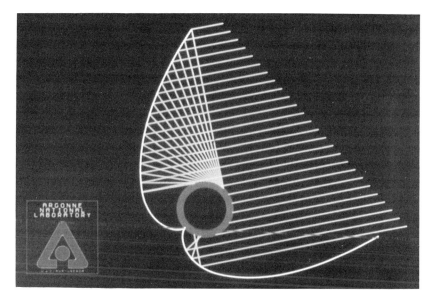

Figure 7.4
Computer generated ray trace of a non-imaging, compound parabolic concentrator (CPC) collector with a concentration ratio of 1.5 shown at an incident angle of 35°.

Roland Winston (see chapter 7 of volume 5 in this series). Computer ray tracing has also been used to analyze the optical performance of reflector-augmented flat plate and tubular collectors as well as multiglazed, multisurfaced, and multisided flat absorbers (see figure 7.4).

There have also been major advances in the ability to analyze the heat transfer processes that occur in solar energy conversion devices. Experiments and powerful computer techniques have been used to improve the ability to calculate natural and forced-convection processes in geometries used in many types of solar collectors. Both convection enhancement and convection suppression techniques have been studied and applied to improve the performance of collectors. These advances are especially important in the development of large-scale solar energy conversion devices, such as central receivers, for which experimental measurements are difficult and expensive.

There has also been a substantial amount of work related to the application of the second law of thermodynamics to solar collectors and solar thermal energy conversion process. Second law analysis is a powerful, but little used, method for optimizing the design of solar thermal conversion process and for matching solar collectors with heat engines (thermodynamic cycles).

7.3.2.3 Performance and Durability Test and Standards

One of the most important accomplishments of the federal and state solar programs was to support the development of standard methods for testing and rating the performance of solar energy conversion equipment. The measurement and reporting of the efficiency of a solar collector or system are not simple tasks. Outdoor conditions are never stationary and are difficult to control; realistic indoor simulations are difficult to produce. Reliable standard test methods are essential, however, because the marketing of equipment is based very heavily on its claimed thermal performance.

Test methods for indoor and outdoor testing of solar collectors and for the testing of some types of thermal energy storage devices were developed and widely accepted early in the program through the efforts of the National Bureau of Standards (later NIST), the American Society of Heating, Refrigeration, and Air-Conditioning Engineers (ASHRAE), and others. Today these methods are used by the independent Solar Rating and Certification Corporation (SRCC) in their widely used OG-100 standard for solar collectors. Methods for testing and rating the thermal performance of solar energy systems were more difficult to develop. Today, a process for the certification of solar water heaters is available, but is only used (required) in a few jurisdictions.

In addition to thermal performance testing, the solar heat technologies program included the development of tests and standards for all manner of important product characteristics such as reliability, durability, compatibility of material, weather and environmental resistance, structural characteristics, and fire safety. Test laboratories were established in different climates and environments to assess the sensitivity of equipment to the local conditions. Standards were established for characteristics such as tolerance of stagnation conditions, rain penetration, and fire resistance.

One of the most important lessons learned in the demonstration field performance programs that were conducted over the past two decades is that system failures are most often traced to improper installation. Failures of non-solar-specific components such as pumps, valves, and controllers also account for more system failures than do those of the major solar components such as collectors (early problems with collectors were usually resolved quickly). These lessons have led to the realization that standards are needed for complete systems and for proper installation. For these reasons, the recently developed SRCC's standard for solar water heating systems, OG-300, includes specifications for all components (and allowable substitutions) and requires installer certification and random inspection of installations.

7.3.2.4 Performance and Quality Control The solar energy industry grew rapidly in the 1970s and early 1980s. At one time there were more than 300 manufacturers of flat plate solar collectors alone. Many were small operations with little technical, manufacturing, or financial expertise. Many of the products they produced were sold because of government incentives to buyers (tax credits and accelerated depreciation), rather than because of the energy savings these products could achieve or their quality, reliability, and durability. In this environment, it is not surprising that not all of the solar products were well designed and well produced. The market and the industry were expanding so rapidly that the usual market feedback tending to assure the emergence of reputable producers and quality products did not have time to play its normal role. As part of its effort to accelerate the adoption of solar energy through R&D, demonstration, and customer information programs, the government played an active role in identifying and helping to correct deficiencies in solar products in the marketplace. Most often the mechanisms were supportive and collaborative, rather than regulatory and punitive; government programs sought to identify technical problems and help manufacturers correct them.

The process of identifying problems worked fairly well. Quality programs conducted by NBS and other national laboratories in support of demonstrations found the major trouble spots fairly early in the program, and the R&D program provided the resource for resolution. Unfortunately, not many of the quality improvements found their way into the market before it collapsed at the end of the tax credits. Those manufacturers who have survived may not have taken advantage of all the technological advances that came from the federally supported R&D, but the market shakeout has certainly reduced some of the quality problems that plagued the earlier market.

The resolution of many of the problems of early solar energy conversion devices really did not require advanced research or the development of exotic new materials. The problems were often the result of failure to apply well-known engineering design principles. Low efficiency of collector modules was often the result of improper optical or thermal design or of mechanical design or production errors. For example, a common problem in early flat plate collectors was improper sealing and venting. This often led to wet insulation or condensate on the underside of the glazing—both of which could cause significant losses of efficiency. Enclosure seals failed for a variety of reasons, including improper selection of materials and inappropriate closure designs. One frequent design error was inadequate allowance for differential thermal expansion. Another was the failure to design for extreme conditions such as stagnation (when the collector is exposed to full solar radiation for a long period of time with no coolant flow or no coolant present), subfreezing temperatures, or wind-driven rain.

Some collector modules suffered from low thermal efficiencies because of improper sizing of the coolant flow passages. This led to poor flow distribution and large temperature variations in the collector absorber. Even when the collector module was properly designed, the system designers often assembled collector arrays with improperly configured or sized piping. It was surprising to some system designers that their collector arrays did not perform as well as the manufacturers of the collector modules claimed. These extra array losses were often the result of poor flow distribution or excessive manifold heat losses either during operation or at night. Additional losses of performance in large arrays of solar collectors were attributed to building or interrow shading, atmospheric turbulence induced by the collector array or adjacent building, or soiling of the collector glazing. Large parasitic power losses (power required to circulate the coolant) were also common in improperly designed collector arrays. In addition to thermal performance losses, improper manifold design also resulted in corrosion, scaling, and freezing problems. Array structural design flaws occasionally resulted in collector or building damage and piping or roof leakage.

It was fortunate that there were some government-sponsored quality programs to identify design and installation flaws because most installed solar energy systems had full conventional energy backup systems and little or no instrumentation. It was not at all uncommon for owners of solar energy systems to be quite happy with systems whose solar energy components were not working and, in some cases, had never worked. This could easily happen when solar energy systems were installed in combination with other energy-conserving measures.

7.3.2.5 Advanced Collector Development The search for more effective means to capture the sun's energy has fascinated people since ancient Grecian times. But the solar era of the 1970s and 1980s provided more opportunity to conceive, analyze, and test new solar collectors than any previous time of history. Not only were more funds available for research, but there were computers capable of analyzing virtually any configuration, new materials with special properties, test laboratories with special equipment for testing solar collectors, and a host of interested engineers and scientists around the world. It is not surprising that a great deal of effort was focused on developing advanced collectors that were more efficient, capable of achieving higher temperatures, or more cost-effective than those known in the first half of the twentieth century. It is surprising, however, that so few of these efforts actually resulted in successful new commercial products.

The greatest government effort by far was, and still is, devoted to the development of advanced collectors for solar thermal power applications. This category of collectors includes distributed and central receiver, line and point focus, single-axis and double-axis tracking collectors. The government programs also included salt gradient solar ponds. The programs for development of these collector technologies were run by the large national laboratories—SNLL, SNLA, JPL, and later, SERI or NREL. Hundreds of advanced concepts were considered, with all types of reflective and refractive concentrators, tracking devices, and receiver configurations. The three concentrating collector technologies that have survived this process and achieved technical, if not commercial, success, are the parabolic trough, the parabolic dish, and the central receiver technologies. All these use reflective optics. Refractive optical systems have been developed for PV applications where the uniformity of illumination they can achieve is important, but have, so far, proved too costly for solar thermal applications.

The advanced technology that attracted the most interest in the commercial sector was the evacuated tube collector. Collector development programs were conducted by several of the major glass and fluorescent light manufacturers, including Owens-Illinois, Corning, General Electric, RCA, Phillips, Sylvania, Westinghouse, Raytheon, and by many smaller companies, with support from DOE. While the development of several types of evacuated collectors was technically successful, that is, initial problems of thermal shock, loss of vacuum, large manifold losses, and so on were largely solved, no U.S. manufacturer was ever able to establish a large enough market to invest capital in the manufacturing processes that were expected to make the product cost-effective. Today, there are no companies manufacturing evacuated collectors in the United States although they are imported from Japan and Ireland. The most successful of the evacuated collectors was, in a sense, the parabolic trough collectors manufactured by Luz for the solar thermal power plants in Southern California. Luz collectors used evacuated receivers with highly selective multilayer coatings capable of operating stably at temperatures near 900°F (500°C). Some arrays even used vacuum-jacketed piping for some of the interconnecting piping.

The CPC and a number of related nonimaging reflector concepts for nontracking collectors were widely developed but never commercialized. The most technically successful of these is the integrated CPC evacuated collector (ICPC), in which the reflector and receiver are enclosed in an evacuated tube (see figure 7.5). This collector requires no tracking but rivals the parabolic trough at temperatures up to 360°–450°F (200°–250°C). Prototypes were built and tested at the University of Chicago, but no manufacturer in the United States has undertaken production. There are reports that one or more Japanese companies plan to commercialize the concept, but nothing is on the market at this time (1996).

Although it was never an official part of the solar thermal heating program, I cannot resist mentioning the most ambitious of the advanced collector development ideas—the construction of large solar thermal or photovoltaic power plants in space, as championed by Peter Glazer and others. In some versions of this idea, mirrors in stationary (geosynchronous) orbits would focus sunlight constantly on an attached energy conversion module (power plant or photoelectric or photo optical device) that would then transmit the power to earth in some different part of the electromagnetic spectrum. In others, mirrors might reflect the sunlight directly to the earth to be converted in terrestrial power plants or simply used to illuminate, or augment illumination of, the earth's surface. Solar reflectors, on orbiting relay stations, have also been proposed to transmit power from one terrestrial power plant to another and, by harnessing the solar wind, to achieve interplanetary voyages in the space age equivalent of sailing ships (the acceleration is low, but very high speeds are eventually possible).

Significant Results of the Federal Solar R&D Programs

Figure 7.5
A CPC collector with one tube in place. The absorber captures virtually all the light entering the trough with the tube installed. Hence the entire trough is black.

In 1993 Russia orbited a reflector that was large enough and accurate enough to increase the luminosity on the earth's surface as it passed over, and apparently did so on a very modest budget. The orbiting mirror is one of the key steps in the Russian program to develop satellite space power stations by 2010, and it was conducted right on schedule. The Japanese also have a space power program with a goal of meeting 30 percent of the nations needs by the year 2050. To the author's knowledge, there are still no formal U.S. development programs for orbital solar power systems for terrestrial needs, but NASA is said to be reexamining the concept. The twenty-first century could well see such concepts given serious consideration here as well as abroad.

7.3.2.6 Cost Reduction R&D Regardless of the application, solar heat can never be economically attractive until the cost of energy delivered by solar collectors is at least comparable to the cost of energy delivered at the same temperature by devices using conventional energy resources. Even at the peak of oil prices in 1980, it was difficult for solar energy to compete with natural gas or coal on a cost per unit energy basis. Thus there has always been an enormous incentive to reduce the cost of energy delivered by solar collectors. The variables that control the cost effectiveness of solar collectors are the cost, the efficiency, and the energy density; because the efficiency is clearly limited and the energy density is not a design variable, much of this effort was directed at reducing the capital and life-cycle cost of the collectors themselves. Much has already been said about the importance of the market size in determining the cost of manufactured (and installed) systems and components. This section discusses the effort to reduce costs through changes in design and materials.

The two basic approaches to initial cost reductions in materials are to (1) reduce the mass of materials and (2) reduce the unit cost of materials. Both approaches were aggressively pursued in R&D programs sponsored by the federal government (and presumably also by private industry). The effort to reduce the mass of material per unit of energy collected took many forms. Structural design changes were especially effective in developing concentrating collectors that were stiffer, stronger, and lighter, and easier to install. Many of these structural design innovations are evident in the modern parabolic troughs, and the stressed membrane reflectors found in today's parabolic dishes and heliostats.

In some collector components, strength or stiffness is not a major consideration; in such cases, effort could be made to replace materials with ones that were simply lighter or thinner. Because flat plate collectors are generally installed on building roofs, they may not require great structural strength of their own. Therefore, one cost-saving approach that received considerable attention in the solar era was the development of flat plate collectors made completely from thin films of strong, lightweight plastics. Plastic films, it was argued, could be laminated together in a continuous process to form the shell, insulation, absorber, and glazing of air- or liquid-cooled flat plate collectors; or they could be substituted for at least some of the heavy and expensive glass or metal components of flat plate collectors. Lightweight collectors have the potential to reduce not only the manufacturing cost of systems but also the installation labor and cost. Major R&D efforts to develop innovative plastic collector products were conducted by Marshall Space Flight Center, Accurex Solar, Fafco Solar, Battell National Laboratory, Brookhaven National Laboratory, and Reynolds Aluminum, among others. Unfortunately, none of these efforts resulted in marketable products. This does not invalidate the approach; under different market conditions, it is quite possible that the development of such lightweight, low-cost collectors would succeed. Low-cost plastics and EPDM absorbers for unglazed collectors for swimming pool and other less demanding applications are commercially successful.

The other primary approach to initial cost reduction in materials is to replace high-cost materials with low-cost materials even if the total mass is increased. For example, the support structure for a parabolic mirror could be made of concrete rather than steel, or the absorber of a flat plate collector could be concrete, sand, wood, or broken beer bottle glass. Such approaches have been even less successful than the lightweight materials approach (see figure 7.6). They tend to have two major disadvantages: the low-cost materials tend not to have the other characteristics needed for good collector performance (e.g., good thermal conductivity for absorbers or poor conductivity for insulators), or they do not lend themselves to high-volume, low-cost production (e.g., a cast concrete mirror structure may be cheaper to build than a single structural steel assembly, but more expensive to mass-produce). There have been exceptions—most notably, salt gradient solar ponds, in which the salt water is the low-cost material used as the absorber.

One approach that has proved quite successful in several types of collectors is to manufacture the collectors in larger module sizes. This can reduce the labor and overhead costs per unit of collector area, and

Figure 7.6
A lightweight collector panel developed by BNL was efficient and easily installed.

thereby increase the collectors' cost effectiveness. This trend has been especially obvious in concentrating collectors, where the module size has increased from, say, 540 ft^2 to 5,400 ft^2 (50 m^2 to 500 m^2) in some parabolic dish and heliostat designs. Similarly, flat plate collectors designed in Europe for large ground-mounted arrays achieve substantial savings using areas of 130 ft^2 (12 m^2), compared to 22 ft^2–44 ft^2 (2–4 m^2) for typical roof-mounted collectors.

7.3.3 Energy Storage

Energy storage is necessary in most building applications because buildings generally require heating, cooling, and hot water both day and night. In addition, buildings often need heat when solar energy is least available: during the winter and cold, cloudy weather. Energy storage is also valuable in power and industrial applications to extend the periods of operation or shift power production to the time of day when it is most valuable.

The predominant type of energy storage employed in solar heating systems is thermal energy storage—either sensible or phase-change. Chemical storage, although an attractive option for many applications, has not yet become cost-effective for most current applications.

Traditionally, solar heating systems for building applications have employed tanks of heated water for energy storage. This is still the most widely used technology, but much effort has been spent to improve its efficiency and reduce the cost. One major source of inefficiency in systems using large water tanks is the mixing of hot and cold regions of the tank due to either natural convection (caused by the buoyancy of water) or forced convection resulting from injection or extraction. A considerable theoretical and experimental effort has gone into research to better understand and control the mixing in tanks, including everything from numerical three-dimensional modeling studies to experimental studies of multiacre salt gradient ponds.

Although water has a remarkable sensible heat capacity, its liquid range (at normal pressures) is too limited for many solar thermal applications. Alternative storage media include a wide range of liquids, some solids, and materials that undergo various types of phase changes. For example, hydrated salts, such as Glauber's salt, have been studied extensively for heating applications in buildings because their transition temperatures are in the proper range and can be tailored to meet the needs of specific applications. Similarly, polymers that undergo structural phase changes have been studied for higher-temperature applications, such as absorption cooling. Ice, of course, has been the storage medium of choice for refrigeration for millennia, and is now used in energy management systems for air-conditioning in large buildings.

Low-cost solid materials are also candidates for energy storage in buildings. Passive heating systems rely on building walls, floors, and even furniture for suppressing the diurnal temperature swing in the conditioned space. Active air heating systems frequently use rock bins for heat storage and occasionally as heat sinks in cooling systems. Large-scale systems for seasonal energy storage may employ in situ rock or earth accessed by natural (aquifers) or manmade heat exchangers (drilled rock or embedded tubing) or water in large earth-coupled excavations, such as rock caverns or earth pits.

Because power and industrial applications usually involve temperatures higher than can be safely and economically allowed in water storage systems, a good deal of research and development has gone into high-temperature storage systems. Liquids with a wide temperature range, such as organic and inorganic oils, can, like water, be used as both the heat transfer fluid and the storage media. Molten salts and liquid metals may also be used as both heat transfer and storage media, but usually present more operational challenges. One of the major successes of the solar thermal power development is the successful field-testing of a molten salt, direct-contact, central receiver, in which the molten salt can flow directly to the thermal storage tank. This system will undergo prototype testing in Solar Two, a central receiver power plant, scheduled for completion in 1995.

7.3.3.1 Design Problems Temporarily Overcome Few thermal energy storage units, other than conventional domestic hot water (DHW) tanks, are standard manufactured products. Most of the installations involve adaptation of products designed for other applications, or the special construction of individual units. Some of the lessons that have been learned by designers and contractors in the implementation of solar thermal energy systems are cited below.

• Prevent high heat losses from piping connections (thermosiphoning) by using backflow restrictors

• Prevent loss of insulation efficiency due to moisture and heat bridges

- Prevent corrosion by using adequate coatings, materials, anode placement, proper antifreeze inhibitors, and so on
- Design tanks to maintain stratification whenever possible
- Choose unpressurized tanks carefully—they are not popular with installers
- Use low-cost materials cautiously; many have restrictive temperature limitations
- Design rock bed systems to achieve proper flow and temperature distributions

Large-scale systems for seasonal or extended storage have received considerable R&D attention in North America and elsewhere. Some of these systems appear to be cost-effective in some solar and nonsolar applications. Aquifer systems have been tested for both high-temperature (>300°F or 150°C) and low-temperature (<300°F or 150°C) applications. High-temperature aquifer tests were technically successful but have not found commercial applications. Low-temperature applications are commercial in some parts of the world (most notably, China) and have attracted commercial interest in the United States.

Other large-scale systems have been mostly investigated in Europe. Rock caverns and drilled rock thermal energy storage systems are near commercialization in Sweden. Systems of piping embedded in natural clay formations appear to be cost-effective for solar and nonsolar applications in several parts of Europe and in the United States. Large earth pits and insulated tanks are still too expensive for most applications.

7.3.3.2 Phase-Change Thermal Energy Storage Systems for Buildings

With the exception of ice storage, which has been very successfully used in commercial buildings to shift air-conditioning load to utility off-peak hours and to reduce the size and cost of air-handling equipment in new installations, phase-change thermal energy storage has not been very successful in building systems. One of the most attractive concepts was to integrate phase-change materials into conventional building materials, such as sheet rock, that could be used to increase the heat storage capacity of passive solar buildings without adding massive elements or unusual architectural features. While some of the phase-change materials developed in the solar program for such applications have been commercialized, they are now found in boot and hand warmers for skiers rather than in building products.

7.3.3.3 Chemical Storage Systems Although a broad range of thermochemical reactions is technically capable of providing efficient storage, and a number have been tried in solar applications, there has been no commercially successful application of thermochemical storage either in the building or power fields. The ultimate chemical storage technology, of course, would be based on hydrogen, and many solar-driven thermochemical processes have been examined, but not commercialized. Many countries now consider the production of hydrogen from renewable energy sources a high priority, and it is likely that the twenty-first century will see this goal achieved.

7.4 Solar Technologies for Buildings

The direct and indirect uses of energy in U.S. buildings account for some 36 percent of the national primary energy budget, and are delivered at a cost of more than $160 billion per year. The thermal energy needed for space heating and cooling and domestic hot water accounts for more than 20 quads of primary energy demand. Most of this demand can be met from heat at the relatively low temperatures delivered by simple solar heating technologies. By maximizing the use of solar energy in low-temperature (low-exergy) applications in buildings, the nation can preserve more of its limited domestic and imported high-exergy fuels for applications that require it. Thus the building sector has been the focus of much of the nation's solar energy research efforts.

Building energy research was conducted in both the solar program and the energy conservation program; this series integrates much of that effort. The integration of building science and architecture is covered in both volume 4, *Fundamentals of Building Energy Dynamics*, edited by laboratory and university researcher Bruce D. Hunn, and volume 9, *Solar Building Architecture*, edited by noted author and publisher of solar and architectural work Bruce Anderson. The volume on building system energy dynamics draws heavily from the work conducted under the energy conservation program, and the volume on building architecture integration examines the work accomplished in both the solar and conservation programs and in the private sector to integrate the results of this research.

Before the DOE came into existence, government programs concentrated heavily on active solar energy systems. Once DOE was established, the solar energy systems and components program for buildings was roughly equally divided between active and passive system work. In the years after 1976, interest in passive systems exploded and the budgets for passive systems grew beyond the active system budgets in 1979 through 1981. The current DOE program for solar buildings is very small, and once again is predominantly an active systems program.

The technical aspects of passive solar technology for heating buildings are primarily treated in volume 7, *Passive Solar Buildings*, edited by a pioneer in passive building science, J. Douglas Balcomb. Techniques and technologies for maintaining building comfort in hot climates are covered in volume 8, *Passive Cooling*, edited by Arizona State University professor and internationally known architect Jeffrey Cook. These two volumes document and detail the many accomplishments and some of the failures of what is considered by many of those who experienced it to be the most fascinating era of solar building evolution. The architectural integration of passive technologies into attractive and livable buildings is covered in volume 9, *Solar Building Architecture*.

All of the major activities in active solar energy systems for buildings are covered in volume 6, *Active Solar Systems*, edited by the solar pioneer and senior spokesman of the solar community George Löf. The accomplishments discussed in this volume, like those in volume 5, are largely a result of the ERDA/DOE R&D programs, but volume 6 also draws from the demonstration and commercialization programs in both the public and private sectors. The use of active solar systems and their integration into the built environment is covered in volume 4, *Fundamentals of Building Energy Dynamics*, and to a lesser extent, in volume 9, *Solar Building Architecture*.

7.5 The Energy Dynamics of Buildings

The designer of solar energy systems for buildings must understand how and when buildings require energy to maintain comfortable temperature and humidity, provide lighting, and service hot water for the occupants.

While it is sometimes possible to separate the design of the building from the design of the building's energy systems, as is traditionally done in the heating, ventilation, and air-conditioning (HVAC) world, it is always more satisfactory for the energy system's designer to be involved in those aspects of building design that determine the heating and cooling loads. In the case of passive solar buildings, the design of the energy system is an integral part of the building design.

While the art of building design has been recognized for millennia as one of our noblest pursuits, building science, and particularly building energy science is relatively new. In order to either minimize the energy required by buildings or take advantage of the sun and the wind to maintain building comfort, engineers and architects needed to know more about the details of the dynamic thermal response of buildings to varying boundary conditions. Thus both the energy conservation and solar energy programs of the early 1970s emphasized the need for a better understanding of building energy dynamics. As a result, both programs sponsored work on experimental and theoretical evaluation of heat and mass flow in buildings, and the development of analytical methods and computer programs to assist the architects and engineers to better design and analyze their buildings.

Perhaps the most important achievement in the building energy analysis field was the development or evolution of large computer codes capable of predicting the behavior of even relatively large and complex buildings. Although these codes are too complex for everyday design use, they are vital to the design of efficient large buildings and provide a more convenient method of validating simplified methods than the experimental approach.

Fundamentals of Building Energy Dynamics reviews not only the large body of work that resulted directly from the government programs, but also the continuing efforts by the professional societies, such as ASHRAE, to provide their members with accurate and up-to-date information. This volume is as much a status report on what is known and how to apply it as it is a compilation of R&D accomplishments. However, in addition to providing a guide for analysis, the volume also contains information on how the national aggregate of buildings use energy, and how the energy use patterns have changed as a result of the energy price shocks of the 1970s and the conservation and solar responses to those events. This information is of interest to everyone in the energy field.

7.6 Passive Heating Systems

Interest in passive solar heating systems grew rapidly on a national scale following the first Passive Solar Conference in Albuquerque in 1976. This was a landmark event, attended by a strange mix of Washington bureaucrats, national laboratory scientists, and hippies. They did not always communicate well. When an architect was asked how much storage the passively heated house he designed contained, he responded that it was ample: two closets in the master bedroom and a full attic. They seemed to spend more time debating what was or was not "passive" than on understanding fundamentals. Although they were sure that passive heating was economical, no one had any idea what the extra costs really were. But there was excitement. Today, people would call what happened a paradigm shift. In a few years all but a handful had given up on active solar space heating and cooling systems, and passive was almost the only game in town.

In the decade that followed the first passive conference (1976–1986), there would be an explosion of public and builder support for passive building design. It is estimated that more than 200,000 single-family residences were built during this time with passive heating features. And not all of these were in Santa Fe! The total number of active space heating or cooling systems installed in residences was probably less than 40,000. Commercial building applications of passive solar designs were not as common, although it is estimated that some 1,500 buildings incorporated passive heating or daylighting features.

But the R&D action was in analysis, evaluation, and promotion of the performance of passive designs. Led by the team at Los Alamos Scientific Laboratory, analysts everywhere were simulating the performance of Trombe walls, sunspaces, roof ponds, and, of course, windows' direct gain. Test boxes and test cells sprang up at laboratories and universities across the country (see figure 7.7). Architects and engineers began cooperating on new designs to exploit the results that were coming out of the laboratories in the form of copious papers, reports, and a series of handbooks. Government design competitions were held, with the winners receiving grants to support construction of the concepts. As more passive homes were built, researchers began to instrument them and study their thermal behavior—at first informally in the communities surrounding the laboratories and universities where interest was high, and later in nationwide programs that included three levels of monitoring.

By the time the passive solar program at DOE hit its peak (1981), the architects and engineers were mostly speaking the same language; the computer printouts had been replaced with less esoteric design guidelines; and rules of thumb and simplified design and analysis tools were developed that architects, builders, and even home buyers could use. Although some of this important work continued, with lagging federal support and a public loss of interest in energy conservation, passive building practices never became standard in the mass housing industry. Most of the 200,000 homes that were built were custom homes that employed passive features because buyers, architects, or builders had both the interest and the means to select the features they wanted. While the custom homes did stir interest because of their distinctive appearance, they were occasionally built with more flair than practicality.

Overuse of direct gain and other features resulted in some buildings becoming overheated at certain times of year, or being too sunny or too exposed for everyday living. The application of passive heating to multi-zone buildings and the understanding of passive thermal communication between zones did not progress as far as passive applications to simpler buildings. The need for simple, preferably passive, means for control of solar gain and loss became apparent, and serious R&D on controllable and switchable glazings became a high priority in the later years of the government passive solar program. Although important advances have been made, there is still no practical, affordable, smart glazing on the horizon.

The possible synergism of combined techniques was recognized but not well understood or optimized. As the movement matured, architects learned to apply moderation in their designs and to achieve building comfort and efficiency through judicious combinations of heat load reduction measures and passive solar features. Had the building industry not fallen into recession and had the cost of gas and oil been less volatile, the average tract home of today might be a product of that design transformation. However, hard data on the actual incremental costs of incorporating passive measures in typical residential buildings are still lacking.

Significant Results of the Federal Solar R&D Programs

Figure 7.7
The evolution of passive solar heating experiments. Test boxes, cells, and buildings at Los Alamos and Ghost Ranch.

Figure 7.7 (continued)

7.7 Passive Cooling

Passive cooling research was the poor stepchild of passive heating. There was never a coordinated program, but individual researchers did receive support and made important contributions. There were also projects that included passive cooling techniques alone or in combination with heating. The biggest concern about cooling in the passive program was how to limit the need for extra cooling in passively heated buildings (i.e., heat avoidance).

The mechanisms available to the architect or engineer for influencing the comfort of a building in a hot climate are the same heat and mass transfer processes that control solar heating systems (and all thermal systems): radiation, conduction, convection, evaporation, and condensation. An important difference is that the driving potential for cooling is much smaller than for heating. The source temperature for passive heating is the temperature of the sun, about 10,260°F (5,700°C), while the sink temperature for cooling is that of the lowest available ambient reservoir, the atmosphere at 80° to 100°F, the earth at 50° to 60°F, or some body of water at a comparable temperature, or the night sky at perhaps −80°F. It is obviously easier to transfer heat from a source at 10,000°F to a building at 72°F than to transfer heat from a building at 72°F to a sink that is often hotter than the building. In fact, the second law of thermodynamics precludes this possibility unless work is done on the system (i.e., mechanical air-conditioning). Thus a passive cooling system must give up heat to the sink at the times when the sink temperature is lower than the building, and must avoid receiving heat from the source and the sink when it is warmer than the building. A similar problem arises in the removal of moisture from the building because the ambient air is often more humid than the building air. The tools that the architect/engineer has to accomplish this herculean task include the size, shape, texture, color, and location of the building surfaces, and the placement of the building itself. As Jeffrey Cook points out in the introductory chapter of volume 8, builders in hot and humid parts of the world have developed many ingenious ways of arranging the elements of buildings to maximize the comfort of the occupants. The emphasis in the U.S. government-supported programs, however, was not only to provide comfort, but to save energy and reduce cost from a building norm that already included efficient mechanical air-conditioning to maintain a comfortable environment.

7.8 Active Systems

An active solar energy system for a building is generally considered to be any collection of water heating or space heating or cooling equipment that includes a solar collector as one of its elements. Conceptually, at least, active solar energy systems differ from conventional building energy systems only in that the conventional heat source is totally or partially replaced with heat from a solar collector. It is not surprising, therefore, that active systems were considered the most direct and straightforward approach to the utilization of solar energy in buildings and received the greatest attention and funding in the government program that evolved in the post–oil embargo years.

Government-supported activities in active systems included R&D, demonstrations, and incentives, for water heating, space cooling, and space heating systems (as well as all combinations of the three). As the solar era started, solar water heating was thought to require the least R&D, and solar cooling the most. Nevertheless, the program began with demonstration and incentive programs of all three applications in addition to R&D. As it turned out, the demonstration and incentive programs were premature.

Solar water heating, although basically a known technology, lacked the infrastructure for widespread deployment or even widespread demonstration. Solar space heating was conceptually simple, but more difficult than expected to implement. Solar cooling had dual obstacles to overcome—the low efficiency of thermally driven cooling processes and the low cost of conventional heat. All three applications require efficient, reliable, and cost-effective solar collectors and energy storage units, but had differing needs for the timing and conditioning of the delivered energy. Thus the progress and success of the active solar program were closely tied to the collector, storage, and materials development activities.

7.8.1 Swimming Pool Heating Systems

The most successful application of the active solar technologies from a market perspective was swimming pool heating. Although this application never received federal incentives, and was not officially included in the DOE R&D programs, solar swimming pool heating has been popular in the sunny parts of the country for decades. Swimming pool heating collectors are normally unglazed and often made from inexpensive polymers. Systems require no storage (the pool is the storage element) and are easy to install and maintain.

According to incomplete EIA statistics, 85 to 100 million square feet of low-temperature collectors have been sold in the United States since 1973. These collectors are used almost exclusively for swimming pool heating. The total shipments to date of low-temperature collectors are about the same as the total shipments of medium-temperature collectors used in domestic water heating and space heating applications. Shipments of low-temperature collectors increased until 1979, fell rapidly from 1980 to 1983 (as the industry concentrated on the sale of water and space heating systems), and have risen slightly to a current plateau of about 6 million ft^2 per year. Most of the sales are in Florida, California, Arizona, and Puerto Rico. In some parts of these markets the level of saturation is fairly high.

7.8.2 Domestic Hot Water Systems

Solar water heating had flourished twice in the United States prior to the beginning of the large government solar programs: in California just after the turn of the century and in Florida in the 1930s–1950s. The government efforts in support of solar water heating therefore focused on demonstration, field monitoring, testing and standards, and public information dissemination. Of course, much of the research and development effort in solar collectors, storage, materials, and control was applicable to solar water heating as well as to the less advanced applications.

The single government program that had the greatest impact on the adoption of solar water heating by consumers was, of course, the residential tax credit program, which originally provided a 30 percent federal tax credit for the purchase of residential solar equipment of up to $2,000 and 20 percent thereafter, with a maximum credit of $2,200. This was replaced in 1980 with a straight 40 percent credit up to a maximum credit of $4,000, which remained in effect until the end of 1985. States added their own tax credits to those of the federal government, so that in some states the net cost of solar water heaters to the residential owner was less than half of the contract price. This was a great incentive for sales, and soon solar water heating companies were everywhere. Their sales forces scoured the residential neighborhoods looking for new sales. Sales skyrocketed, reaching a level of about 11 to 12 million square feet of collectors in 1981–1984. In all, some 80 million square feet of solar domestic hot water (SDHW) collectors may have been installed, most of it during the pre-1986 tax credit era. The most positive view of this burst of activity is that it could have produced total energy savings of perhaps 0.2 quad over the life of the systems, assuming the SDHW systems delivered about 0.2 MBtu/ft^2 per year (2 GJ/m^2 per year) and lasted about ten years.

From the vantage point of nearly ten years of hindsight, however, it is possible to see the tax credit impact in less favorable terms. When the residential tax credits were eliminated in 1985, the shipments of medium-temperature solar collectors dropped from nearly 12 million to barely 1 million ft^2 and 90 percent of the manufacturers and installers turned to other businesses. The market had been built primarily on the basis of the availability of tax credits, not the provision of energy services or energy savings, and could not be sustained without them. The industry and the sales force that had been mobilized to sell tax-credit-subsidized systems did not always place a high priority on quality of products and installations that could deliver energy savings year after year.

Although the marketing and incentives programs for solar water heating technology received the most attention and were the most controversial, there were simultaneous programs and activities in R&D. While most of the technology development was carried out by private industry with little direct support from government, the government was active in monitoring the advances and provided valuable feedback to the fledgling solar water heating industry. Among the most important of the government programs were those establishing standards for testing and rating of solar water heating systems and components discussed above. In parallel, government programs at NBS, Argonne National Laboratory, Marshall Space Flight Center, the RSECs, and elsewhere, tested and monitored the performance of systems, components, and materials in the field. The feedback from these activities to industry and the government R&D programs helped to improve performance of the industry's products and guide research along pragmatic lines.

The performance and reliability problems that plagued some of the early systems were solved by reducing design and installation errors. Installation errors were found to be the most common reason for failures or poor performance in the field; most of these errors could be attributed to market growth that exceeded the ability of the infrastructure to accommodate it. Early systems did suffer from some poor solar components or system design flaws, but most of these were quickly corrected. The more persistent problems were attributed to some of the non-solar-specific components such as pumps and controllers, and to the aforementioned problems with improper installation. The emerging codes and standards, such as SRCC's OG-300 standard for solar water heating systems, therefore not only specify standards for all components and restrict substitutions, but also provide strict guidelines for installation and for random inspection of systems in the field. However, as a practical matter, these measures either were not adhered to or came too late to save the industry from the collapse that followed the end of solar tax credits.

Among the technological advances made by private industry, both during and since the solar era, has been the adoption of a greater diversity of design of solar water heating systems. The earlier booms in California and Florida generally relied on simple designs and design refinements made by one or more innovators. During the 1970s and 1980s system designs proliferated. Those making lasting marks on the progress of the technology were integrated passive thermosiphon systems that require no pumps or controls, integrated collector and storage (ICS) systems that can withstand short periods of freezing weather, low-flow systems that maintain storage tank stratification and reduce plumbing and pumping costs, self-pumping and photovoltaic pumped systems that are self-controlling and do not require power from the electric grid, and more reliable drain-back systems to prevent freezing in cold climates.

7.8.3 Space Heating Systems

Unlike solar water heating or swimming pool heating, active solar energy systems for building space heating were rarely used in the United States prior to the energy crisis of 1973. Until then, there were only four known solar houses in the United States: two research houses at MIT, one at the University of Delaware, and the one built in Denver by *Active Solar Systems* editor George Löf. Because space heating is such a large part of the U.S. energy budget, the government solar energy plans developed in the 1970s placed considerable emphasis on both R&D and demonstration programs for solar space heating.

The government demonstration programs and, later, the tax credits stimulated a demand for active solar space heating systems that eventually resulted in the installation of perhaps 40,000 residential and 1,000 commercial systems. The peak year was 1984 when 2.4 million square feet of medium-temperature solar collectors were used for space heating. The total installed area of collectors for space heating is not known, but could be between 10 and 20 million ft^2. Assuming 15 million ft^2 with a productivity of 0.1 MBtu/ft^2 per year (1 GJ/m^2 per year) and a life of twenty years, solar space heating could contribute a total of about 0.03 quad to the nation's energy needs.

The installation of solar heating systems also came to a virtual standstill in 1985 with the expiration of the residential tax credits. Solar heating showed some signs of recovery early in the 1990s when concerns about the environment and the security of the oil supply raised public awareness of energy issues once again, but the recovery has not been sustained.

Solar space heating presents a greater challenge for the system designer than swimming pool or domestic water heating because of the greater mismatch between the demand and the availability of solar energy. In addition to the diurnal cycle, solar space heating systems must contend with the seasonal variations in both supply and demand. This has proven to be very difficult, especially in climates where cold weather is also cloudy. If a system is designed to meet a major portion of the winter heat load, it will have excess capacity in the summer and poor annual utilization of the investment in solar collectors. If it is designed for full utilization of the summer resource, by including, for example, the water heating load, as almost all system designs do, it will not contribute to the winter heating load. Thus all compromises will result in poor utilization of collector area relative to water heating alone.

Two alternatives to resolving this dilemma have been pursued. In the United States, designers sought to combine solar space heating with solar-driven space cooling (air-conditioning) in order to achieve higher utilization of the collectors. This approach was relatively unsuccessful primarily because the efficiency of the cooling equipment that can operate at temperatures compatible with solar heating is so low. These combined systems were also often so complicated that they did not work very well.

The other solution to the solar/load mismatch is seasonal energy storage, which was vigorously pursued in some of the northern European countries and included, to a lesser extent, in the R&D programs in North America. The technical advantage of a system with seasonal storage is that the collector array can be optimized for full utilization of the available energy, thus maximizing the energy available per dollar invested in collectors. The technical trade-off is the increased size and efficiency of the energy storage system needed to assure a continuous supply of heat through the winter. Phase-change thermal energy storage or chemical energy storage, could in principle, be used for individual buildings, but no practical and economical systems of this type have been developed. Systems storing sensible heat in water, soil, or rock can be used for seasonal heat storage but, to be economical, must be quite large (large enough for the equivalent of a few hundred homes), and therefore are applicable only to central systems for large buildings or district heating. Systems of this type have been successfully demonstrated in Europe and are nearing commercial readiness in countries where the use of community district heating is common, such as Sweden and Germany. Only a few small proof-of-concept projects have been built in North America, but the approach could become more popular as the utility sector explores new energy services and new business opportunities.

Aside from the basic economics of solar space heating, the technical issues associated with space heating were also more difficult to solve than anticipated. Technical success was demonstrated in some installations, but many more proved to be disappointing. Solar heating systems turned out to be much more demanding in terms of design skill, installation practice, and control system functionality than the conventional systems they augmented or replaced. Most of the companies designing and installing solar heating systems did not do enough business to ascend the learning curve. Monitoring showed that the best-performing systems were those in which designers had a continuing role in the monitoring and evaluation of the project, were thus able to recognize and correct their initial mistakes.

Air heating systems' requirements for handling and distribution systems were especially hard to meet in the market. Technical success was achieved, but great care was required. Solar-augmented heat pumps (SAHPs) failed to offer cost advantages; series SAHPs saved more energy than parallel systems but cost too much. Direct systems (no storage) enjoyed some success for limited applications such as industrial make-up air preheating.

7.8.4 Space Cooling and Heating Systems

Just as the main drawback to the use of solar energy for space heating is the mismatch between solar availability and heat demand, the main appeal of solar-driven space cooling is the coincidence between supply and demand. This appeal is heightened by the growing worldwide demand for air-conditioning, and the rising cost of providing electricity to meet the summer peaks of many utilities. But like solar heating, the goal is illusive. In order to produce a cooling effect, solar heat must be used to drive a heat engine (a thermodynamic cycle). Unfortunately, the efficiency of available heat engines and cycles at temperatures compatible with solar collectors is not very high. This means that large collector areas are required to meet given loads. A major part of the government solar cooling program was devoted to improving the efficiency of the three energy conversion processes available for space cooling: the Rankine cycle, absorption, and desiccant cooling. This research was successful in increasing cycle efficiencies to some degree. Unfortunately, this just produced a moving target for solar cooling systems, since the improved efficiency machines could just as easily use conventional heat sources (usually natural gas) rather than solar heat. At the same time the efficiency of standard electrically driven vapor compression chillers and heat pumps have also been increasing.

With oil and gas prices at or near all-time lows in real terms, the prospects for solar cooling do not look favorable in the near term. As things now stand, for solar cooling to be competitive in the marketplace, the first requirement is that solar collectors be capable of delivering energy to heat-driven cooling machines at a cost less than or equal to the cost of supplying heat from natural gas or oil. Except in a few places where petroleum products and electricity are very expensive (such as some of the Caribbean islands), this is a difficult requirement to meet.

While advanced cooling system research is certainly important to the efficient utilization of energy resources of all kinds, the goal of achieving greater utilization of solar energy would be best served by those R&D and market development activities that result in more cost-effective solar collectors regardless of their application. This overriding goal has apparently been recognized at the federal energy budget level, where the amount appropriated for solar cooling research has been reduced to zero.

7.9 Solar Architecture and Planning

One of the important lessons architects and builders of passive homes have learned is not to use the more extreme passive techniques and to use direct gain in moderation. By limiting the solar savings fractions to 25–30 percent, they can make the building look and feel quite conventional. At this performance level, standard building practice seems to provide enough thermal mass to meet diurnal energy storage needs. This is good because adding thermal mass has turned out to be more expensive than originally thought. The other important reason for settling for a modest energy saving is that more aggressive strategies can lead to summer overheating or air-conditioning energy requirements that offset winter heating energy savings. Daylighting is also tricky. It is true that in commercial buildings, daylighting may reduce cooling requirements arising from artificial lighting, but it may also increase the cooling load from direct gain during some of the year, or may add to the heating load.

One explanation for the modest success of passive techniques in practice is that the loss of interest in energy saving of the 1980s occurred before many of the advances of the 1970s could be converted to practice. Another explanation is that better control of the heat gains and losses may be needed to achieve really efficient buildings. The promises of controllable glazing, angle-selective glazing, and spectral-selective glazing and roofing systems have not yet been realized in the market.

7.10 Solar Thermal Technologies for Industry and Power

As mentioned in chapter 2, our series was originally intended to cover power and industrial applications of solar thermal technologies in two additional volumes, which, for reasons beyond the control of the editor-in-chief or the MIT Press, were never completed. The technologies involved in power and industrial solar systems, however, are covered to some degree in volume 5, *Collectors, Energy Storage, and Materials*, and in volume 10, *Implementation of Solar Thermal Technology*. The two planned volumes, *Fundamentals of Concentrating Systems* and *Distributed and Central Receiving Systems*, would have greatly amplified on the technology and R&D results of the government programs and would have covered the many important projects and demonstrations in detail. Fortunately, there has been greater continuity in the solar thermal power and industry programs than there was in the buildings program. Although there were budget reductions and serious setbacks in both the public and private sectors, the programs still exist, continue to build on the research and lessons of the 1970s and 1980s, and possess some programmatic memory. Repositories of the report literature exist in the government laboratories that still have active programs (SNLA and NREL). Figure 7.8 shows a historical perspective of some of the R&D and demonstration accomplishments of the solar thermal program from 1976 through 1985. In addition, there have been other publications that have documented the R&D accomplishments of the solar decade.[1]

7.10.1 Power

In spite of an impressive outlay of government funds for research and development on solar thermal power options by a number of countries ($902.7 million in the United States alone through 1990), commercial development is limited to a single company and a single utility. Luz International built nine solar electric generating systems (SEGSs) in southern California that now provide 354 MW of power to the Southern California Edison (SCE) grid.

Although the commercial status of solar thermal power is very much in doubt at this time, the commercial market that did exist had a profound effect on the state of the technology. Very important advances were made that establish new benchmarks for the solar industry as a whole.

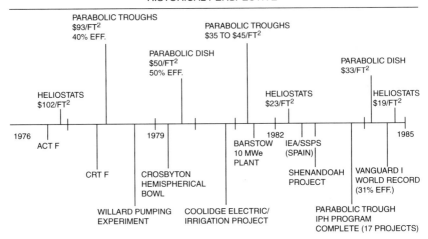

Figure 7.8
Historical perspective of the accomplishments of the Solar Thermal Program.

In 1984, when the first Luz plant, SEGS I, was completed, the cost of power from the plant was estimated by Luz at 24 cents/kWh. According to Luz, the cost of power from SEGS IX, the 80 MW plant completed in 1990, was 8 cents/kWh. Luz had initiated a new technology development program in 1989 that would have brought the cost of power from SEGS XII, scheduled for completion in 1993 but never built, down to 5 or 6 cents/kWh. These cost reductions were brought about by a combination of technological improvements, economies of scale, and fiscal concessions from suppliers and investors based on performance. They were, not coincidentally, necessary for the survival of the company, for at the same time the government incentives were systematically eliminated or greatly reduced, and the price of natural gas, to which the price paid by SCE was tied, fell by 50 percent. At the target cost of 6 cents/kWh in 1993, the new plants would have required no further incentives or subsidies, but in 1990 when costs were 8 cents/kWh and avoided costs payments were tied to natural gas, which was at a twenty-year low in real dollars, the few remaining tax credits and exemptions from the federal and state governments were needed to survive.

The technology employed in SEGS IX is based on parabolic trough LS3 collectors using back-surfaced silvered glass mirrors, highly selective

Figure 7.9
Solar electric generating systems (SEGS) at Kramer Junction in Southern California.

absorber tubes enclosed in evacuated jackets, and organic heat transfer fluids with evaporators and superheaters to drive the conventional steam cycle turbines (see figure 7.9). Natural gas provides the backup energy that allowed the plants to operate during cloudy periods and to extend the hours of operation during the utility's peak demand periods (this had a large positive impact on the plant's revenues). The LS4 collectors under development for SEGS XII were to have front-silvered mirrors, enhanced selective surface coating, direct steam generation and superheat in the receiver, and north-south tilt of 8 degrees. LS3 collectors operated at a peak temperature of 700°F (390°C) and drove the power block at an efficiency of 37.6 percent. LS4 would have operated at a peak temperature of 820°F (455°C) and provided a 4.1 percent increase in turbine efficiency. (The optical efficiency and $F_R U_L$ values for LS3 and LS4 are 0.80 and 0.32 W/m^{2}K, and 0.83 and 0.25 W/m^{2}K, respectively.)

The progression of the cost of the installed collector array for the SEGS plants is shown in figure 7.10. As noted before the installed cost per unit

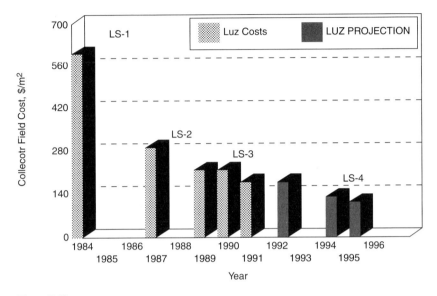

Figure 7.10
History of installed collector field prices for SEGS plants and projections for future plants.
Source: Renewable Energy Consultants, Inc.

area of these high-quality, high-temperature collectors is now lower than that of any glazed flat plate collector on the market today. In most sunny locations, if these collectors could be installed at the same cost as the SEGS plants, they could also deliver low-temperature heat at energy costs that are far less than flat plate collectors.

The Department of Energy believes that both central receiver and parabolic dish collector technologies have greater potential for the solar thermal electric market than the parabolic trough. In fact, DOE phased out research and development support for troughs in 1982 and never supported the Luz development. The cost goals used by DOE for central receiver and dish power are 5 cents/kWh. To achieve these goals, central receiver plants in the southwestern United States will probably have to be 200 MW or more in capacity. A parabolic dish power plant, on the other hand, will have integral heat engines and turbines and could provide local power at a plant size of 25–50 kW. The small module size and high efficiency of the dish system could be a big advantage for remote areas where power costs are high and total demand is small. Prototype dish concentrators with integrated Stirling cycle drive generators have achieved a record solar-to-electrical conversion efficiency of 31 percent.

The largest investment of government funds in solar thermal power was in the development of central receiver technology. The showpiece of the program was the 10 MW Solar One power plant in Barstow, California. Built as a joint venture between the Department of Energy and Southern California Edison, Solar One operated from 1982 to 1990 and produced more than 27 GWh of electricity. The plant is now being refitted with a new molten-salt, direct-contact receiver, and was recommissioned in 1996 at an increased power level of 20 MW. In addition to the pilot plants at Barstow, the government built a large central receiver test facility at Sandia National Laboratory in Albuquerque that has been in operation since 1977, and supported a variety of heliostat and receiver development projects in the national laboratories and private companies.

7.10.2 Industry

In spite of substantial government-led efforts in the United States, Germany, Japan, and other countries, solar technology has made practically no penetration into the industrial process heat market. The reasons are more financial than technical. Because most industries are able to write off fuel costs as business expenses, the value of energy savings is diminished; capital investment in energy production or savings equipment can result in higher property taxes; capital depreciation rules may not be favorable; and on and on. But the most important reason is the high expectation for return on investment that many industries use to make investment decisions. The rate of return (ROR) requirement is usually in the range of 20 to 30 percent (although stockholders will tell you that few industries achieve such earnings). Thus investments in solar heating plants for industrial process heat are seldom undertaken.

The technical status of solar industrial process heat technology is defined by the technology for solar thermal power discussed above. The parabolic trough collector technology developed by Luz could deliver heat at temperatures up to 820°F (455°C) at costs that would be attractive at discount rates used by utilities, but probably not with industry ROR requirements.

There is one potential industrial application of solar technology where the special properties of solar radiation add value to the thermal energy. Research has shown that moderately concentrated solar radiation can catalyze the thermal decomposition of some chemical toxins. With the current international concern for toxic waste management, it appears that these photocatalytic processes may find significant commercial applications.

7.11 Implementation of Solar Thermal Technologies

The role played by government in the development of solar thermal technology is discussed at length in this volume. If one agrees that government attempts to act in the best interests of society, then its actions are often required to achieve societal goals or to achieve them more rapidly and economically than would otherwise be the case. There are, however, many questions about the proper roles of government and the private sector in promoting those societal goals. The view of many in the 1970s was that the development of alternative energy resources was such an urgent national priority that virtually any role was appropriate if it accelerated the achievement of this goal. Among the many alternative energy technologies, solar energy was considered to offer substantial near-term opportunities, and some applications of solar energy were thought to be technically mature enough to quickly enter the market place.

There was, therefore, broad popular and political support for government funding of activities that would hasten the commercial development of solar energy technologies. Under less urgent conditions, some of these activities would have occurred naturally in the private sector. In the Carter years these activities were referred to as "commercialization," and included a broad range of industry, market, and information support programs. When the Reagan administration took over in 1981, its first priority was to eliminate all commercialization programs from the Department of Energy (actually its first priority was to eliminate the department itself) based partly on the belief that "commercialization" was an inappropriate role for government. The commercialization programs represented a large fraction of the total budget for solar heat technologies (Ronal W. Larson and Ronald E. West estimate that the total cost, including outlays and forgone revenue, of the commercialization program was about 1.5 times the outlay for the R&D programs), but the reasons for their termination were philosophical, not budgetary.

Programs representing more than half of the government's efforts to develop solar heat technologies could not be left out of the solar documentation project that led to this series, but to include them was the greatest challenge faced by the project. Not only were the commercialization programs the first to be dismantled, but even the word "commercialization" was placed on the list of forbidden words by DOE officials. Thus the volume that covers the solar heat technologies commercialization work (volume 10) came to be called *Implementation of Solar Thermal Technology.*

Editors Ronal W. Larson and Ronald E. West turned out to be an ideal team for the task of recruiting authors and editing this volume. Although an electrical engineer by training, Larson was already deeply involved in solar policy issues in 1973 and 1974 when, as IEEE Fellow, he served as staff scientist to the House Committee on Science and Technology and worked on the first two solar energy bills passed by Congress. From 1974 to 1975, he was with OTA and served as program manager for OTA's first full assessment of the solar energy program. West (also coeditor of volume 3, *Economic Analysis of Solar Thermal Energy Systems*, in this series) is a professor of Chemical Engineering at Colorado University with a long history of publications and objective analysis of solar programs and projects. Together, they possessed the knowledge, objectivity, and editorial skills to complete a story that, at the onset of the program, many believed could not be told. Time, of course, was both an ally and an adversary. As the fervor of political change weakened, it was easier to write objectively, but more difficult to reconstruct what was actually accomplished. It is this author's hope that the discussions, evaluations, and insights offered in *Implementation of Solar Thermal Technologies* may help guide the actions of future efforts to accelerate the development and commercialization of solar technologies and other technologies.

The remarks in the following sections regarding the success or appropriateness of the various aspects of DOE's commercialization efforts, however, are the present author's opinion and do not necessarily reflect those of volume 10's authors or editors.

7.11.1 Demonstration and Construction Programs

If initial R&D shows success, demonstration becomes an essential part of technology development and may be an important precursor to commercialization. However, if we learned anything from the government solar programs of the 1970s, it was that you have to be very careful when, where, and how a technology is demonstrated. The ill-conceived heating and cooling demonstration program of the mid-1970s only demonstrated how poorly prepared the solar energy industry was to deliver workable heating and cooling systems. It was not only technically premature, but it failed to sufficiently involve the industry or the customers in the financial success of the projects. The same could be said of the industrial process heat demonstration projects. The utility power demonstration projects, of which Solar One was the most visible, were better planned in that they did involve customers in a substantial way. They probably should be judged successful tests of concepts, although they fell far short of demonstrating the technical and economic feasibility of solar thermal power generation. That was nearly demonstrated by the privately developed solar power plants built in response to the provisions of PURPA and the generous power purchase agreements required by the California Public Utility Commission (Standard Offer 4 in particular).

The Solar Federal Buildings Program was the largest and most visible of the construction programs. Some $50 million was spent for this ambitious program to promote the adoption of solar heating and cooling technologies through installations on government buildings of all types, with the ultimate goal of providing a growing market for the solar industry. Some 700 solar energy systems were installed, but the program did little to open up the federal market to solar technologies. The program had many problems, including poor performance of many of the installed systems, but perhaps the most telling was that the host agencies' local offices, which installed the systems, usually received none of the energy savings achieved through the use of solar energy (or other energy efficiency measures). Because, as a rule, any savings in energy operating budgets reverted back to the home agencies, the managers in the field took little interest in the projects, other than to comply with the regulations. On the positive side, the systems were monitored and subjected to performance and design reviews. In this process, a great deal was learned that could make the installation of future government and commercial solar projects more successful. An excellent series of design and installation guidelines were issued at the end of the program, in cooperation with ASHRAE. Unfortunately, because of the low price of competing energy forms, the market for solar heat technologies was already dead by the time these guidelines were published.

In the author's opinion, if the resources devoted to the Solar in Federal Buildings Program and to the Residential and Commercial Demonstration Programs had been focused on creating a sustainable market in selected parts of the country instead of being spread so thinly over the nation that nothing really took hold, the industry and the use of solar would be far ahead today. Unfortunately, political considerations were an important part of the demonstration programs, and thus the programs were not capable of such a focus. Although the private sector is less likely to make the same mistakes, we see today in the solar water heating industry a tendency to try to open new markets rather than to increase the market share in the best current markets.

It is pointed out repeatedly in this and other volumes that to be successful, the solar thermal industry needs to develop the capability to deliver heat from solar collectors as cost effectively as heat from conventional fuels. To do this, it needs a viable, sustainable market, although the market does not have to be national, and it does not have to encompass all applications and all companies. Local markets for the best applications in the best solar regions should be stimulated and allowed to grow and diffuse to less desirable regions as they become more competitive. What the country (and the solar thermal industry) needs, but has never achieved, is at least *one* successful, cost-competitive, supplier of solar collectors that can be used in multiple applications.

7.11.2 Quality Assurance Programs

The two aspects of quality assurance that received the most attention were (1) the development of codes and standards for products and services; and (2) the enforcement of those standards, as well as other programs designed to protect consumers from poor products and unethical practices.

The development of standards has already been discussed in the sections above. Although a vital part of the solar heat technologies program that was, we hope, ultimately successful, it came too late to avert many of the problems associated with poor products and poor installations. Developing standards for a new industry is a difficult and lengthy process under the best of conditions. Adopting and enforcing those standards in codes is even more difficult. It is not surprising that the process met with limited success given the short period available for it to develop.

Customer protection was probably an impossible job from the onset. The idea that millions of end users of energy could be educated to select and install sophisticated energy conversion equipment that was safe, reliable, and cost-effective in less than a generation was just not realistic. Nor could they be protected from their own ignorance.

The problem is that solar thermal technology was, and still is, being sold to the wrong customers. In the present author's opinion, if the wide adoption of solar energy for building applications is ever to occur, some party will have to be responsible for the application of solar energy to those energy services needed by the end users. That is, some party other than the end user must purchase, install, and maintain the energy conversion equipment and derive its income from the fee charged for the energy services delivered. Whether that party will be the public and private utilities as we now know them, building owners, or third-party energy service companies, remains to be seen. However, someone with the motivation (profit), technical capability, financial resources, and public trust must take the responsibility for the selection of reliable, efficient, cost-effective equipment, and its continuing operation.

7.11.3 Information Programs

There were three main aspects of the government information programs: public information, technical information, and education and training. Much of the government-supported information programs were targeted at the general public. These resulted in the outpouring of huge numbers of publications, pamphlets, and articles covering all aspects of solar energy and its applications. It was considered essential for the public to understand and accept solar energy technologies, because most of the technological approaches were marketed to the end user. These programs were judged to be relatively successful at producing and disseminating information, but not successful in convincing the public to adopt solar energy conversion technologies. As noted above, this may have been a hopeless cause. Because the end-user market collapsed with the fall of oil and gas prices and the end of tax credits, as well as the general perception of poor performance of the solar energy systems, the information programs may have to be repeated when conditions are again favorable for solar thermal applications. Hopefully, the next time, less involvement will be required of the end user.

The technical information programs were established to prompt the rapid exchange of information between industry, academia, and the various government R&D activities. These programs were needed initially because the usual channels, professional societies, trade journals, and conferences were not ready to accommodate the rapid explosion of interest and technical data that came from the new solar programs. The technical information programs and their special reviews of technologies were relatively successful in reaching and influencing their target audiences, and some of them continue today. The authors of this volume certainly hope that the information program that led to the publication of the present series will be considered worthwhile and successful.

If the goals of the solar programs were to be achieved, there would have to be a huge transformation of the infrastructure to support the new technologies. Scientists, engineers, technicians, production workers, installers, salespersons, and entrepreneurs would all have to be trained or retrained. Government programs were established for all these groups, but, with the demise of the solar industry, the need was short-lived. Many who trained for years for a career in solar energy have been disappointed.

7.11.4 Technology Transfer Programs

It is sometimes difficult to distinguish between information dissemination and technology transfer. The government laboratories conducted R&D programs and made results available to everyone through the information dissemination programs. There were also some examples of cooperative programs between government laboratories and individual companies to develop new products or processes, most often materials. Most of those involved in this process from the government laboratory end saw this as a successful process, and volume 10 cites a number of examples involving different laboratories.

The private sector view of technology transfer, as volume 10 points out, was somewhat different. Some viewed the laboratories as real or potential competitors for research and development support and as present or future suppliers of services for the commercial sector. Many worried that their ideas would be taken over by the laboratories and either developed or transferred to competitors, or that government-provided services, such as testing or consulting, competed unfairly with the private sector. As one who has been on both the laboratory and private side of the issue, the present author believes there is some merit in these concerns. It is difficult to strike a fair balance in trying to advance technologies, support the legitimate needs of the private sector, and not pose a threat to some companies or individuals in the private sector.

7.11.5 Incentive Programs

Incentive programs were considered necessary to accelerate the adoption of solar energy technologies that were in the national interest. Without incentives, it was felt, conversion would be delayed too long and become more costly and even dangerous to energy security. Incentives might be needed to develop a sustainable market, but should not be necessary to sustain it. The government established a number of financial and nonfinancial incentives for the solar programs. The financial incentives included the tax credits, the Solar Bank, and the grant programs. Each of these is covered in volume 10.

In addition, the government provided disincentives for solar and other renewable energy technologies in the form of continuing subsidies for established energy providers. The Alliance to Save Energy (ASE) estimated that in 1993, government energy subsidies amounted to about $36 billion annually, of which less than 1 percent could be counted as incentives for the use of solar energy. Some analysts have placed the total value of subsidies accorded oil, including the costs of international security, in the neighborhood of $100 per barrel. Whatever the figure, the goal of a level playing field for all players in the energy business is far from a reality.

Of the financial incentive programs provided for solar thermal technologies, only the commercial and residential tax credits were large enough and lasted long enough to have any significant impact on the development of a market. Although their long-term impact on the development of the industry they were intended to establish has been described before in negative terms in the present volume, in the short term, they did achieve the limited objective of increasing the sales of solar equipment and creating jobs. Unfortunately, too many of the jobs they created were for door-to-door salespersons, not factory workers and technicians. Whether the total economic value of the solar tax credits exceeded their cost has never been determined.

Not only was the market for solar energy systems unsustainable without the tax credits, but the energy savings the systems were intended to achieve were often unsustainable as well. Systems that were badly designed or installed did not deliver the savings they should have, and the system owners generally had no way of knowing if their systems were even operating. As installers went out of business, owners were left without anyone to turn to for maintenance and repair. At the same time, the costs (real) of natural gas and electricity were dropping and home owners were much less concerned with energy availability and cost. As a result of poor installation and poor maintenance and repair, it is likely that the actual total energy savings have been far less than the 0.2 quad that could have been realized if all systems installed had performed as expected.

However, in the present author's opinion, the failure of the government incentives to produce a sustainable market and industry should not be interpreted as a failure of the technology, nor even a condemnation of the economics of solar water heating. The technology is capable of delivering long-term performance, and with the proper market delivery mechanisms, it can compete on even terms with conventional water heating in some parts of the country. The main problem, as we are just beginning to acknowledge, was that the products were being sold in the wrong markets and the incentives were designed to produce sales and did not assure continued performance. The most recent extension of the commercial tax credits for renewable energy technologies corrects that latter error and bases the credit on the energy actually delivered. The next step is to realize that solar technology can be marketed to utilities or third-party energy services that can profitably sell the thermal energy service (not products) to the end user. This market delivery mechanism has numerous advantages, but what is most important, it makes the service provider, rather than the end user, responsible for the efficiency and reliability of the energy conversion equipment.

Because of the end-user marketing strategy, most of the market action was in the residential sector. Commercial and multifamily markets proved disappointing because the commercial end users either could not benefit from energy savings or required high rates of return on energy-saving investments. Apartment owners generally passed costs through to a renter and had little knowledge or interest in energy conversion technologies. Except for a brief period in California when third-party systems were popular with investors seeking tax shelters, solar water heating has been excluded from the substantial rental sector of the residential market.

The experience with the residential and commercial tax credit incentives suggests to the author that these incentives should be based on results, that is, the delivery of energy from renewable resources, or the reduction of the use of energy from nonrenewable resources, not on the sale of equipment. Hardly anyone disputes that the tax credits were a major cause of market abuses. The new tax credits for wind and biomass power are based on energy delivered. It remains to be seen how successful these will be. Even if the approach is 100 percent correct, they could fail because the incentive level is not set correctly or timed properly. It is clear that the incentives should also be tied to the cost and availability of conventional energy resources, but the proper relationship is difficult to predict.

The nonfinancial incentives provided for solar thermal power through the PURPA legislation, which required utilities to purchase electricity from third-party providers and gave some preferences (and restrictions) to renewable energy resources, when combined with the federal and state tax credits and exclusions, could have been successful. If the present volume had been completed in 1991, it would have said that Luz's accomplishment is the first real success story in the solar industry and demonstrates that solar energy can succeed in the marketplace with the proper assistance at the right time. It would have pointed out how properly structured and timed government and utility incentives can help a fledgling industry compete successfully on a playing field otherwise badly tilted in favor of the established energy technologies. Unfortunately, in July 1991, Luz was forced to file for bankruptcy and liquidate its assets. Although this development has not affected the operation of the nine existing plants, which are owned by investors, it has brought to a halt, at least temporarily, construction of new plants that had been proceeding at a rate of 80 MW per year and were expected to increase and expand into northern California and Nevada before the end of the century. There are potential successors to Luz who may decide to purchase rights to technology and the utility contracts, but if this option does not materialize soon, much of what Luz accomplished technologically will be lost, and it may be a long time before anyone will be willing to risk the treacherous game in which the yearly success or failure of an industry is in the hands of legislators who have little interest in or knowledge of that industry and its problems.

Now all that can be said is that the federal government and California, after bringing the solar thermal power industry to the brink of success, allowed it to fall back into an abyss. The Congress and the governor of California did this through procrastination, not malice, but the effect was the same. Delays in approving an extension of small, but still vital, government incentives were the final straw that led already nervous potential investors in SEGS X, the tenth solar electric generating system scheduled for construction in 1991, to withhold capital so long that Luz could not meet its cash flow requirements and had to close its doors. The incentives that made it possible for Luz to establish itself and advance its technology no longer exist, and it may be a decade before "business as usual" energy economics will make it attractive for even the most risk-tolerant companies to enter the business in a substantial way.

7.11.6 Organizational Support Programs

Volume 10 cites, as organizational support activities, programs that involved the utilities, state and local governments, international cooperation, and labor, law, or environmental issues. Surprisingly, the environmental issues received little attention in the early development of the solar heat technologies program. Most of the emphasis was on energy conservation and security. Today the linkage between energy consumption and the environment is the critical issue driving energy policy, at least until the next energy crisis.

In this author's view, the involvement, or noninvolvement, of the utilities was the most important failure of the solar thermal commercialization program. The program certainly recognized the need to involve the electric utilities in the programs for the development of solar thermal power, but failed to include the utilities in an important way in the commercialization of solar heating and cooling. Although this was consistent with the public attitude toward utilities in the early 1970s, it was not a winning strategy. Had a conscientious effort been made to involve the utilities in the solar commercialization program in the beginning, the incentive programs could have been quite different, perhaps rewarding performance rather than sales. Instead, the feeling in the country was that the utilities were profiteers who would only turn the solar movement to their selfish interests, who would try to "meter the sun." Some utilities did undertake solar programs, but in many states they were barred from participation as owners or providers of energy services.

The international programs for commercialization of solar energy have been more prominent in the photovoltaics field than in solar thermal technologies. Solar thermal technologies have been included in programs such as the Committee on Renewable Energy Commerce and Trade (CORECT) and earlier trade missions, but usually the emphasis has been on PV. There have also been a number of international collaborative research programs conducted under the auspices of the International

Energy Agency or by means of bilateral agreements. Although these activities were not usually commercialization activities, they are discussed in volume 10. At one time the international research programs were regarded as giveaway programs and were mostly terminated or restrained by the Reagan administration. Now that the situation is reversed, and the R&D centers of some solar technologies, such as solar thermal building technologies, have moved abroad, one wonders why the United States does not take greater advantage of the opportunities for international cooperation to improve our technology base.

Note

1. Carl-Jochen Winter, Lorin L. Vant-Hull, and Rudolf Sizmann, eds., *Solar Thermal Research and Technology* (Spring, 1991); Thomas B. Johansson, Henry Kelley, Amulay K. N. Reddy, and Robert H. Williams, *Renewable Energy Sources for Fuels and Electricity* (Washington, DC: Island Press, 1993); International Energy Agency, *Renewable Sources of Energy* (OECD/IEA, 1987).

8 Observations and Lessons Learned

Charles A. Bankston, Donald A. Beattie, and Frederick H. Morse

As promised in chapter 1, we have ended this volume with our observations and lessons learned during the course of our energy program experience. Readers of the other volumes will find specific lessons learned relating to the various technical aspects of solar heat technologies, and volume 10 includes an "Evaluation Discussion" of the demonstration program that is a compendium of lessons learned from that program. The following discussion, however, in keeping with the wide perspective of volume 1, looks at some of the broader issues. It examines the propriety of government involvement in energy, and it discusses policies and practices that affected the success or failure of the government-sponsored solar energy RD&D program in the context of other energy programs. Although the acronyms R&D and RD&D have been used somewhat interchangeably, both in this and earlier chapters, not all government energy R&D programs lead to the second D, "demonstration."

8.1 Why a Government Role in Energy?

The energy sector represents a small, but critical, fraction of the U.S. economy. In 1973 we saw the impact that even a small shortfall in the energy supply could have on the economy and the quality of life of the nation. The United States and most of the developed world have enjoyed many decades of cheap and plentiful energy—based largely on low-priced oil; indeed, our economies, transportation systems, and life-styles have become heavily dependent on low-priced energy. We know that the earth's resources of easily converted forms of energy, such as fossil fuels and fissionable isotopes, are currently being consumed at rates that will result in their effective depletion within one or two centuries. There is also evidence that our excessive conversion of some of these stored energy resources to heat and power has environmental consequences that put the long-term habitability of our planet in question. Today, governments must face their responsibilities to assure energy security and environmental stability for current and future generations.

The authors' generation of Americans witnessed profound swings in public and government attitudes toward the use of energy and the security of its supply. From the early post–World War II years, when energy supply was entrusted to large multinational corporations (including the Seven Sisters—Exxon, Mobil, Texaco, Gulf, Chevron, Shell, and British Petroleum), to Carter's declaration that a secure energy supply was "the moral equivalent of war," to the Reagan administration's "hands-off, the market-will-decide" approach, to Desert Storm, we have seen our government's role undergo significant shifts.

Until 1973, the energy marketplace, as exemplified by oil, was stable and prices were low. This stability was maintained by the nearly total dominance of production and refining capacity by vertically integrated, multinational corporations. These companies controlled every step of the oil energy business from exploration to the gasoline pump. Although they competed with one another and with coal and gas producers in the consumer fuels business, they did not compete for the price of crude oil at the wellhead. Production levels were determined by consumer demand for the end products (gasoline, fuel oil, etc.). This stable market era ended in 1973 when OPEC nations, through nationalization of production and increased market share, gained control of production levels. The price of oil quadrupled in just over two months.

The response of the free world to this loss of control over oil price and production, along with dire forecasts of imminent oil and gas resource depletion based on faulty models of the resource base as well as demand growth, was to question the ability of the marketplace both to manage the remaining resources and to adopt renewable energy supplies in time to avoid serious energy curtailment and economic collapse. The fear that the world was quickly running out of oil led governments—especially in the industrialized countries—to take unprecedented measures to assure continuity of their energy supplies. That the dire forecasts turned out to be wrong, that our understanding of the response of world markets was too limited and the government-sponsored programs, designed to meet a crisis that has not yet materialized, never achieved their goals does not, however, mean that the programs themselves were wrong. One of the lessons to be learned from the events of the 1970s and 1980s is that decision makers need forecasts and data based on more reliable and more robust models. Another is that the market does work to optimize energy use, but its lag time is long and it plays by its own rules, not necessarily society's. The value the market places on energy, that is, its market price, does not include the full societal costs of its use and, as a result, will not lead to a societally optimal mix of energy usage. Had the market in the mid-1970s priced energy to include the full societal costs of energy, some of the renewable energy technologies (e.g., passive solar buildings, solar water heating) would have been more successful in gaining significant market shares.

Corporate America is ruled by the "bottom line"—quarterly profits. With this mentality, it is difficult for the U.S. market, as presently structured, to move from established energy practices to those responsive to a broader spectrum of considerations. But perhaps the most profound problem with the market is pricing. The true societal costs of end-use energy from a particular resource should include the cost of discovery, production, transportation (including infrastructure development and maintenance), and refining; the cost of capital, operation, and decommissioning of the energy conversion equipment; the cost of maintaining secure and fair access to the resource; the cost of environmental or ecological impact resulting from extraction, transportation, conversion, and disposal; and the cost of loss of life or health attributable to its use. The idea of incorporating societal costs into the price of products responsible for those costs would not be unique to the energy field. Through excise taxes, we already incorporate some of the societal costs of tobacco and alcohol into their prices. Other products, such as buildings, have societal costs that are not now fully embedded in their market prices. But energy, as ubiquitous as it is, should be a major target for price reform.

The authors believe that a proper role of government is to set policies that will encourage free market forces to minimize the total long-term societal costs of maintaining a reliable and acceptable supply of energy required for the social and economic well-being of its citizens. How would society make the transition from an economy in which energy is highly subsidized to one in which energy users pay the full cost? Clearly, a rapid transition would be disruptive, even perilous to some businesses or countries, and would be strongly resisted. But because the transition is ultimately necessary, it makes sense to begin before there is a crisis.

8.2 Government Energy Policies: What Will Work?

The U.S. government mounted a multifaceted response to the oil embargo and the oil price increases of 1973 that included massive programs for alternative fuel supplies as well as accelerated RD&D and commercialization of renewable energy sources. The projections of the day were based on both erroneous assessments of the accessible reserves of oil and gas and unrealistic, price-independent rates of growth of energy demand. Many were convinced that the energy reserves were being depleted too rapidly to trust in the free market to make the transition to more secure, renewable energy supplies. Thus, it was argued, the government should provide incentives to accelerate adoption of energy conservation practices, renewable energy, and energy-efficient technologies that would benefit societal goals. This approach included the unsuccessful establishment of residential and commercial tax credits for solar heat technologies.

We now know that these incentives were not only unsuccessful in creating a sustainable market for solar energy technologies, but they were also harmful to the very industries they were intended to promote. In the authors' views, there were three major errors in the design of the incentive programs intended to promote the use of solar energy in the 1970s. The most important was that the incentives were for the purchase of equipment and not for the delivery of energy. We have already discussed the consequences of that error in chapter 7. The second was that the incentives were provided to the end users who knew little about the technologies and were not motivated to call for changes to the available equipment. Finally, the incentives were not tied to the market prices of alternative energy sources or the societal costs using any rigorous methodology. Neither were they timed to begin when the technologies were ready to enter the market, nor to phase out as entry was successfully achieved. If these mistakes are to be avoided in the future, it will be necessary to develop more rigorous methodologies for determining how and when to provide incentives for desirable change—and when to impose disincentives to continued harmful practices.

It is admittedly difficult to conceive of a rational, fair, and easily implemented method of delivering incentives to encourage the adoption of solar heat technologies. In most cases, delivered energy is not measured for an installed solar energy system. Faced with the problem of determining the true energy benefits of installing the 1970s state-of-the-art solar energy systems, it is easy to see why Congress succumbed to the simplicity of residential and commercial tax credits. But the simplistic tax credits did not accomplish their objectives; more sophisticated and rigorous incentive strategies are required.

Because the ultimate goal should be to transform the energy market into one in which transactions are based on total societal costs, the first step toward designing workable policies is to agree on a rational methodology for estimating the total societal costs of energy produced from different sources. In the last decade, many countries have come to realize the importance of including these costs in making decisions or policies on energy matters. Although there is no generally accepted definition or methodology for calculating the total costs, in 1989 the American Solar Energy Society Round Table on Societal Costs, estimated the total extra cost of energy production—including such societal costs as atmospheric corrosion of equipment and structures, crop losses and soil contamination, impacts on public health, disposal of radioactive waste, military forces needed to maintain energy security, subsidies for all forms of energy, and job losses of all sorts—at from 100 to 260 billion dollars per year. That is equivalent to an additional cost of 2 cents/kWh of electricity or 25 cents/gallon of gasoline on average.

A number of studies in the late 1980s and early 1990s attempted to calculate the societal energy cost by fuel and production type. A comparison of some of these studies done by the authors in 1993 again showed a wide range of methods, terms, and values. For conventional coal-burning power plants, for example, the average of seven studies was 8.5 cents/kWh and the range was from 0.6 to 30.3 cents/kWh; the averages for oil and gas were 3.9 cents and 1.6 cents/kWh, respectively. Most of these studies were based only on environmental impact and did not attempt to include other factors such as security, subsidies, and jobs.

Although there is considerable uncertainty in the values to assign to the various terms in the cost equation, and which terms should be included once the procedure is established, the uncertainty in the values will gradually be reduced. The process should not be fundamentally more difficult than maintaining the economic indicators that heavily influence the national economy today. There is no reason why the societal cost of energy conversion and other industrial and commercial activities cannot be tracked in the same way we now track the consumer price index.

Once the societal costs of energy from all sources are established, it will be easy to determine not only which commercial energy technologies are too expensive and should be discouraged, but also which emerging energy technologies are ready for market entry and warrant some form of incentive, and which need further R&D or demonstration. New technologies that can be applied at costs lower than the societal costs of conventional energy supplies they would replace are beneficial and should be encouraged, even though their market price may be higher than that of the commercial energy technologies they would replace. Technologies that show potential for near-term societal cost competitiveness may need further development or demonstration, and promising technologies whose ultimate costs cannot be determined may need more research and development. However, because the market does not yet use societal costs as the basis for investment, even those technologies with low societal costs may not make any headway in the present market without some intervention. Government incentives may still be needed to bring technologies with lower societal costs into the energy mix and to close the gap between market prices and societal costs.

There are basically three ways our government can influence markets where it is not a major participant: (1) it can use tax credits or other fiscal incentives to lower the effective prices of desirable alternatives; (2) it can use taxes or other fiscal disincentives to raise the prices of undesirable alternatives; and (3) it can command all the various media channels to convey its message to the citizenry. Because they will apply to only a small segment of the energy economy, substantial incentives for switching to renewable energy forms may be offered with minor perturbation of the economic system—at least initially. Incentives, however, further exacerbate the problem of fairness in paying the societal costs of energy conversion by adding energy-related taxes to the general tax burden. The practical argument against disincentives is that, applied against dominant energy resources, they would have to be very large to achieve an appreciable switch to the use of renewable energy. Indeed, most economists estimating the price increases necessary to motivate energy consumers to switch from conventional to sustainable energy resources based on price elasticities would conclude the costs are too great for the economy to bear. However, sudden imposition of such large disincentives on conventional energy conversion is not needed.

The authors believe that a combination of incentives and disincentives, dynamically linked to each other and to the total societal costs of energy through a rigorous methodology, applied over a long period of time can move the energy market to a more sustainable structure. This can be accomplished gradually, without any serious disruption of business or economies, a way that slowly lowers the total costs of energy to society, and never exceeds the societal costs of "business as usual." Our concept has three fundamental principles:

1. Provide subsidies (incentives) only for conversion and delivery of end-use energy from renewable resources when it is possible to deliver such energy at societal costs lower than those of the conventional forms of energy they would replace;

2. Provide these subsidies to qualified suppliers of renewable energy in an amount proportional to the difference between the societal costs and the market prices of the conventional forms of energy replaced, and fund them from taxes (disincentives) on the conventional forms of energy just sufficient to cover the cost of the subsidies; and

3. Adjust the subsidies and taxes periodically (every few years, say) so that transition is achieved, the subsidies are reduced and eventually eliminated when no longer needed, and the taxes on conventional forms of energy are eventually transferred from funding the subsidies to paying for the external costs of the forms of energy being replaced (e.g., use part of the tax on coal to pay for the black lung program).

The implementation of this concept is not simple. It requires long-term commitment and fiscal discipline by government and patience on the part of those who advocate change, but if faithfully applied, these three principles would assure a market-driven transition to a sustainable energy economy at minimum societal cost.

For example, subsidies offered to businesses or utilities that supply heat or power to end users on the basis of principles 1 and 2 would serve as powerful incentives for any renewable energy technologies that could deliver energy at costs lower than the social costs. They would also be powerful incentives for businesses and utilities to supply end-use energy services to customers unable or unwilling to make energy-efficient choices for themselves. Initially, the cost of the subsidies could be offset by small carbon or Btu taxes collected from the suppliers of the conventional energy forms being replaced. The amount of the taxes would be determined by the costs of the subsidies provided. Because, initially, the market shares of the subsidized forms of energy would be very small and the market shares of the taxed energy forms large, the taxes would be very small. As the use of renewable energy increased, the tax rates would need to grow, further increasing the prices of the conventional energy forms relative to the subsidized energy forms, and thus accelerating market penetration of the renewable energy forms. As the renewable energy market shares increased, (1) the subsidies could be cut back in such a way as to maintain a fixed relation between the market prices of the conventional and renewable forms of energy; and (2) economies of scale would cause the unit cost of delivering the renewable energy forms to fall, thereby further decreasing the required subsidies.

Although the taxes on nonrenewable energy forms could be reduced as the need for renewable energy subsidies diminished due to both cost reductions and market saturation, it might be wise to continue to collect the taxes on nonrenewable energy forms and to apply any excess revenues to pay for the external costs of those energy forms, societal costs not included in energy prices (infrastructure, defense, environmental restoration, health, etc.). To the extent that renewable energy forms also contributed to such societal costs, they might also be taxed. At the same time, the government should start to pass on to the energy suppliers involved other societal costs of energy conversion and environmental remediation. Citizens and business now pay some of the societal costs of energy conversion through income taxes (defense, infrastructure, environmental restoration, etc.), medical costs and insurance rates, and tangible and intangible losses in quality of life, while other societal costs are passed on to future generations. Correcting the inequities in our energy pricing system will take a long time and great political courage; eventually, the prices of all forms of energy would be brought into line with their true societal costs. Note that this scheme never increases the total societal costs of energy for the total population. It simply shifts some of the burden from the general population to those who use the forms of energy that increase the societal cost. The increased costs of some forms of energy would be offset by decreased taxes and nonenergy costs. This may seem utopian, but it is fair, equitable, and farsighted—and perhaps even politically feasible. There are many indications that our political system is becoming less willing to pass on liabilities like the national debt, social security, and radioactive waste storage to our grandchildren.

Although Germany and Denmark have recently imposed small conventional energy taxes to cover the cost of their subsidies to renewable energy suppliers, to the authors' knowledge, no country has yet developed the rigorous methodology needed to set and maintain incentives and disincentives in accordance with societal costs.

8.3 Role of Government in Energy RD&D

What is the proper role of government in energy RD&D? Let us focus primarily on the RD&D process as a whole, classic R&D and demonstration, recognizing at the outset that there is probably no simple answer to the question, and that the DOE renewable energy program, which served as our background, was somewhat atypical.

Where the government is the sole or primary customer, such as has been the case for defense technology or throughout much of the history of our space program, strong arguments can be made that government should lead the RD&D efforts, both in funding and performing the actual R&D at national laboratories. Even in these two RD&D arenas, however, the private sector conducts the lion's share of federally funded RD&D in its own facilities and manages many of the national laboratories under government contract. With the proposed privatization and commercialization of some space capabilities and the increasing importance to the defense industry of overseas weapons sales, the "government as primary customer" role has become blurred. And in the Industry R&D program, the government leverages its funds and permits industry to be more responsive to downstream government needs, further reinforcing the importance of the private sector. For defense and space, demonstration (fly before you buy) is the equivalent of providing a near-commercial product to the customer, in this case the government, because it leads directly to major government buys and utilization.

Is there an analogy between defense and space RD&D and energy RD&D? The answer, unfortunately, is yes and no. Federal funding for energy supply technologies RD&D, as we have seen in the earlier chapters, has a long history. If we include in the federal energy role other activities such as tax incentives and regulation, the government has been a controlling influence as both funder and doer, certainly in the last fifty years. In this respect, then, the analogy is good, especially for nuclear energy RD&D. However, if we look at the government as energy technology customer, the analogy is bad. Although a major customer, the government is neither the sole nor even the dominant customer. Because the government fails this last test, strong arguments were made during the Reagan administrations against the value of government participation in any energy RD&D program, except basic research.

To rationalize and consolidate government's role in energy RD&D in the mid-1970s, a period of perceived national crisis caused by widespread concern over temporary energy shortages, the finiteness of fossil energy resources, and the need for secure supply, ERDA was formed. There was strong logic for this move because the government's programs, right or wrong, were highly fragmented, and the tradition of government leadership in nuclear energy RD&D was deeply ingrained in the thinking of most decision makers. ERDA's objective and guiding philosophy was that the government, with the help of industry, would develop on a pilot scale a number of new energy supply technologies that industry could not or would not develop on their own, and put them "on the shelf" until the nation (read "energy industry") needed to use them. In this philosophy, the demonstration phase (pilot plant) concluded government's role; commercialization would be left to the private sector. This was a new and interesting concept proposed by a pro–private sector, Republican administration, and never before attempted on such a large scale.

ERDA never had the opportunity to carry out its objective. In ERDA's short, two-plus years of existence, good progress was made in bringing several technologies toward pilot plant demonstration, but events beyond its control overtook the programs. With hindsight, one could easily argue that the objective would never have been achieved, or if achieved, that the technologies probably would have languished "on the shelf" through obsolescence or lack of industry interest. Furthermore, government managers lacked the necessary background and understanding to properly respond to the economics of a highly complex marketplace, the driving force for the application of new energy technologies.

The elections of 1976 changed the energy field completely. With the new energy philosophers brought to Washington by the Carter administration, government's role was to be greatly expanded for a sector of the economy that is, arguably, the most complex. This complexity is created because energy encompasses all segments of government at all levels, the private sector at all levels, and the public at large. From taxes to regulation, federal power marketing, federally owned dams, federally owned oil reserves, the growing strategic petroleum reserve, public and private utilities, interstate pipelines, large and small oil and gas producers, refiners and distributors, builders and developers, to the gas station on the corner, the energy sector is involved in all aspects of our life. To place all these varied activities under the purview of one huge government bureaucracy, DOE, was a monumental undertaking, the magnitude of which was not fully recognized by its initiators. Once again, the perceived crisis resulted in an action undertaken with bipartisan support that would have long-lasting consequences.

DOE, first proposed by President Ford and established by President Carter, attempted to break new ground, in particular by intervening in the private energy sector through price regulation, mandatory purchasing agreements, expanding tax incentives and accelerated commercialization of new energy supply technologies in addition to traditional support of RD&D. These actions built on a legacy of activist initiatives supported by a series of sweeping laws enacted by the 93rd and 94th Congresses and resulted in many positive accomplishments, especially in the field of renewable energy. Most of the policies implemented by the Carter administration were short-lived and had little long-term impact; instead, the problems they created reinforced the perception that government intervention in the energy sector was a mistake. Most of the actions taken during the Carter administration were overturned by its successor.

The Reagan administration, in reaction to the Carter policies, and hearkening back to the philosophical approach of the early Nixon years, espoused limited government participation in only high-risk, cutting edge R&D that the private sector would be reluctant to undertake or unable to afford, and began to phase out all demonstration programs. RD&D budgets at DOE were slashed, especially for renewable energy, and notice was given that DOE itself would be eliminated. But once again, even this constrained and flawed definition of government's role proved untenable. Congress refused to cooperate. Industry was of two minds. While gradually reducing in-house research during the past twenty years, industry still conducts some high-risk R&D, especially where potential end products would be critical to long-term, competitive positions. However, the energy industry looked to the government to subsidize energy RD&D. And for renewable energy technologies, some form of "leveling the playing field" with conventional energy supplies was considered essential by Congress because solar energy systems were still not economically competitive without it. The Reagan administration took the position that there was no shortage of conventional energy and, therefore, the marketplace would sort out any supply inequities. As it turned out, this position was correct: the actual price of energy fell over the next eight years.

The Reagan administration approach, primarily deregulation, led to lower energy prices in the short term but did not address the longer-term nagging problems of what if the demand for oil continued uncontrolled and the OPEC cartel and other foreign oil and gas producers were to reduce supply to importing nations. The approach also did not take into account the growing environmental concerns over burning fossil fuels and utilizing nuclear energy. With a Democratic Congress controlling the purse strings and favoring a more active government role, an uneasy truce resulted. DOE survived, and the constrained definition of government's role in energy RD&D eased. Government's role in commercialization, however, was eliminated, and the support of regulations, tax incentives, and other forms of government intervention espoused by the previous administration also disappeared. Tax incentives for the solar industry, for example, were permitted to lapse.

But inconsistencies continued as to which supply technologies would be favored under this constrained approach. Following the administration's lead, DOE's R&D budgets in the 1980s for conventional energy (oil and gas, coal, and nuclear energy) once again became predominant. Solar energy took a back seat, and the playing field tilted once more in favor of conventional energy. There was, of course, an immediate impact on the solar industry; sales decreased and many companies went out of business. The bright spot was that energy conservation, or the efficient use of energy, which was often closely allied to renewables, remained an accepted practice by industry and consumers, even without large DOE budgets attempting to influence the marketplace. The bottom line still controlled: money could be saved through adoption of energy efficiency techniques. Government's role in furthering energy conservation is marked by both successes and failures. Corporate Average Fuel Economy (CAFE) standards, for example, were a success; energy performance standards for buildings were not. The impact of government RD&D on conservation is hard to assess; conservation RD&D budgets were never large. Highlighting the benefits of energy conservation through public awareness announcements linked to government programs was certainly important. Mandated, automobile fuel efficiency is probably where government actions had the greatest positive effect in transforming the marketplace, although government RD&D did not supply the technology that made this possible. The public at large got the message and expected the auto industry to market fuel-efficient vehicles.

How does government RD&D affect the energy marketplace? First, it is important to recognize that changes in how the economy uses different types of energy are evolutionary, not revolutionary. The widespread adoption of each new energy source, from wood burning, to coal, to oil and gas, took place over periods of fifty years or more. Accomplishing the widespread adoption of nuclear energy in only twenty years is perhaps the only government success story, depending on your outlook. Some government-sponsored nuclear energy programs such as the terminated breeder reactor and fusion energy have not come close to application, in spite of more than twenty years of large-scale, government-funded RD&D. Synthetic fuels and oil shale are two other examples of failed government programs; billions of taxpayer dollars went into these technologies, thus far with no contributions to the nation's energy economy. And although major progress was made in reducing solar energy systems costs via the government-funded solar or renewable energy programs, the overall impact of solar technologies on the marketplace has been small. The lesson is, even though the fifty-year cycle is not complete, it is hard for government-supported, energy-related RD&D to make a major impact, even with strong private sector involvement.

The authors firmly believe there is a role for government in energy RD&D. The question is, how should it be pursued? Although the jury is still out on its success, and renewable energy advocates might object to its promotion, DOE's Clean Coal Program may be the best example, to date, of how government and industry can work together effectively on energy technology. Underway for the past seven years, the Clean Coal Program is aimed at demonstrating advanced technologies that will result in the more efficient and cleaner burning of the nation's largest fossil fuel resource. The utility industry is now building or retrofitting power plants based on the technology developed and, in addition, beginning to take the technology overseas, providing a double dividend of new business and reduced environmental impact for energy-starved countries. The cost of the program to taxpayers has been approximately $2.58 billion, with the private sector contributing approximately 65 percent of the cost of the demonstrations. That industry has been willing to ante up its own funds to the tune of more than $4.5 billion would seem to confirm their belief that something worthwhile will come from this partnership. Certainly industry cost-sharing has been higher in this program than in any other energy demonstration program.

Would these advances have been made without the government participation? Probably not, certainly not as quickly. Are the results worth the dollars spent? Again, the jury is out, but the prospects are encouraging. Early renewable energy demonstrations typically required little or no industry cost-sharing, which undoubtedly contributed to the lack of success in the commercialization of renewable energy technologies. If one agrees that the Clean Coal Program is an example of the successful pursuit of new energy technology by the government, how did it happen? The answer is partnership—government funds and expertise matched by the private sector, driving toward goals mutually agreed on, not dictated by the government. In recent renewable energy programs, this lesson has been learned. Both the photovoltaic manufacturing automation technology program and the Utility Photovoltaic Group demonstration program are based on strong government-industry partnerships.

Such an approach is not new. The most successful of such R&D partnerships was that begun in 1915 between the government's National Advisory Council on Aeronautics (NACA; later the foundation of NASA) and the nascent aircraft industry. Starting on a small scale, the NACA programs evolved into major programs and national laboratories with state-of-the-art facilities and bright, cooperative staffs. Everyone in the aircraft industry would agree that the results of that partnership enabled U.S. industry to take and maintain the lead in aviation technology and aircraft sales worldwide.

Although the first successful heavier-than-air flight in 1903 was accomplished by the Wright brothers without government support, many European countries were vigorously pursuing this technology at a national level. After the Wright brothers' flight, most of the technology advances in the next fifteen years were made in Europe, from better engines to better materials and airframe design. The NACA-industry partnership reversed this trend, and within twenty years, U.S. industry was ahead to stay.[1] The recently passed General Agreement on Tariffs and Trade (GATT) will undoubtedly have an effect on how such future partnerships will be structured, but our foreign trade competitors may nevertheless find a way to continue their long-established close government-industry relationships. To stay competitive, not just in the energy area, continuing partnerships modeled after the Clean Coal Program and NACA/NASA must be maintained and improved upon to assure successful investments in RD&D. At the very least, if the government is to sponsor energy

RD&D, the public and Congress must be prepared to set their sights on the long haul.

Most would accept government involvement in basic and applied research as justified and appropriate. Even the Reagan administration, while attempting to impose across-the-board draconian cuts in government-sponsored R&D made an exception of these areas. Future advances in energy technology will be driven by new discoveries in materials science. All energy systems improvements, regardless of fuel or energy source, will be dependent on finding less costly materials by which a Btu or kilowatt can be delivered to the end user. Materials research is an especially attractive target for government-sponsored R&D, and perhaps the most cost-effective. Such research requires dedicated, sophisticated laboratory facilities and personnel. These resources exist today in our national laboratories. The taxpayers' investment in these laboratories, accumulated over decades, should be realized by assigning them the primary mission of conducting materials research. This research should be performed with industry in a carefully structured collaborative program driven by industry's perception of need.

Demonstration is probably the most difficult phase of RD&D to define in terms of the proper government role. Because pilot or prototype demonstrations to prove a new technology, although important, are usually costly and can also lead to dead ends, this critical phase must be carefully planned and executed. When industry determines a technology is ready for demonstration, the best approach will be to let it provide the motivation and take the lead supplying the majority of resources, as we have seen for the Clean Coal Program. Depending on the type of demonstration, the government role could differ markedly and should be jointly agreed upon by all the stakeholders. Inevitably, government-industry RD&D partnerships will be required to bring new energy technologies to the marketplace, not just renewable energy technologies. How government responds to that need should be predicated on the many lessons learned, some documented here.

Finally, a few thoughts on the government's fitful role in commercialization of renewable energy technologies. In the 1930s only the government could have undertaken the building of the many high dams for hydroelectric power—the single, successful example of government "commercialization" of renewable energy on a large scale. And in the 1950s only the government could have taken the lead in advancing nuclear technology to the point where utilities could bring nuclear reactors onto the grid.

Using these two examples as a guide, government programs initiated in the late 1970s to support commercialization of renewable energy technologies seemed logical and useful in spite of obvious differences in the technology and applications. Renewable energy projects, by their very nature, lend themselves to relatively small-scale, dispersed applications. For solar heating and cooling, we are often looking at single buildings or small groups of buildings. Solar thermal plants seem to scale best at a size of a few hundred megawatts. It follows that, although capital costs may be high for renewable energy, per Btu or kilowatt delivered, the overall project costs are small when compared to typical, central power generation. Thus, when a renewable energy technology is competitive in all aspects for a given application, financing and management can easily be undertaken by the private sector. The government's role in commercialization should be to move the energy market toward prices based on total societal costs of energy, as previously discussed in section 8.2.

In view of the above, the authors continue to support government-industry partnerships for renewable energy RD&D; we believe that government has a critical policy role in rationalizing energy prices, and an important, but limited, role in the purchase of commercial systems, based on total societal costs. The maintenance of effective government-industry partnerships in RD&D requires *continuity* of goals, direction, and management. Energy research requires skilled managers and researchers, resources, and most importantly, time. As our national research capabilities are presently structured, support of energy research by the national laboratories will be instrumental in bringing new technologies to the fore. Their facilities and technical staffs are not duplicated in the private sector; all that is needed is a clear-cut mission. Coordination and dissemination of research results to all interested parties, restricted only by proprietary rights, is another important and perhaps unique government function that will speed the pace of development. Eventually, development will require a vision of future markets and a commitment to making a profit through long-term provision of quality products and services, best done in the private sector, with technical and financial support of government.

8.4 Management Changes and Shifting Program Goals

Many government programs experience significant changes in direction or emphasis with every change of administration, especially when a different political party occupies the White House. But changes also occur even when the same party succeeds to executive power but under a new president.

Of all the government-sponsored R&D programs, those dealing with energy seemed to be the most affected by the changes of administration. Without discussing here the major differences in energy philosophy between Republican and Democratic administrations, and the shifts in philosophy that occurred even within a given administration, suffice it to say that each change of administration provoked a rethinking of energy policy and R&D goals. As mentioned in earlier chapters, these changes were especially onerous in the field of renewable energy. Support for R&D programs in renewable energy vacillated between almost zero to "the sky's the limit." R&D programs for other energy sources (fossil fuels and nuclear), although experiencing some swings in support, were always able to maintain more level funding. In addition, energy programs were also afflicted by the introduction of completely new organizational entities in 1975 (ERDA) and in 1977 (DOE), and by the expressed desire to abolish one of those entities (DOE) in 1980. What were the impacts of such instability on solar programs? They were manifold.

From 1975 to 1989, the top government manager for solar programs changed twelve times, on average every fifteen months (Teem, Hirsch, Beattie, Walden, Savitz, Stelson, DeGeorge, Tribble, Collins, Fitzpatrick, Berg, and Davis). The first "rule" in Washington programs is that with each change in program leadership comes an organizational change, and this was true, to some degree, for each of the top solar energy manager changes. The result: shifting program responsibilities, personnel transfers from program to program, in some cases, personnel departures and loss of "corporate memory," and often, demoralization of the key government resource, skilled managers. In some instances, reorganization was needed and productive based on the changes in program emphasis. In other cases, the organizational changes brought only disarray because the changes were so ill conceived and short-lived that the staff involved never had a chance to understand and rationalize their new jobs. It is doubtful any private sector company could have survived such turmoil. And although there are obvious differences between the private sector and government programs, there is a common thread: good management requires management stability.

Beginning in 1975 with the establishment of ERDA, the top solar/renewable energy manager became a political appointee, with a few interim "Acting Managers," or recess appointees, who did not have the imprimatur of a Senate confirmation. These managers brought different backgrounds to the job, ranging from strong science and engineering experience and skills in managing large R&D programs, to none of the above. And of course, the political appointees were obligated to support the political agenda of "their" administration. The authors recognize that a political agenda is an inherent part of any government program, but also believe that R&D programs, by their very nature, cannot succeed in a climate of constant change precipitated by political tugs of war. Great amounts of money are wasted or poorly spent when the R&D cycle, once started, is not permitted to achieve a logical end point, within a reasonable time frame. Such waste occurred in most of the solar energy programs.

Are there ways to minimize waste of R&D program resources? Part of the solution would be multiyear funding. If one can assume that a national priority and goal can be approved for a given energy R&D program through some "neutral consensus mechanism," then multiyear funding in support of multiyear program commitments would permit rational programs to be designed and completed. Such funding would not dispense with careful program review and oversight but would put increased emphasis on program management by the government agency responsible for assuring that objectives and schedules were being met.

A second part of the solution is to bring stability and competence to high-level R&D management. The best way to achieve this would be to reduce the number of political appointees in agencies charged with the responsibility of conducting R&D and instead fill the top positions with career managers of proven experience. Depoliticizing top management positions would greatly reduce the size of the army of lower-grade political appointees who find a home with each new administration, thus removing another source of program instability. Congressional oversight should be exercised at an appropriate level. Congress must recognize that its efforts to micromanage R&D programs will lead to program distortions, disruptions, and in some cases, program failure, and that it must resist the urge to intervene unless mismanagement can be shown. Congressional staff have a long history of such "micromanagement," rarely with positive results. Congress and new administrations should also recognize that programs, once started, especially large programs, are difficult to modify or terminate. This fact, again, argues for consensus and careful planning.

Several commissions have looked into the problems cited above, but their recommendations have largely been ignored. It is past time to face these problems and solve them.

8.5 Needed Procurement Changes

Our final comments have to do with the procurement of government-supported RD&D projects. We have made a case for research in the national laboratories, joint development, and industry-led demonstrations. There are good technical and policy reasons for these recommendations, but to be successful we need to minimize what government does worst—R&D procurement. This is a problem that is endemic to all government programs, but the solar heat technologies program is a prime example of how devastatingly inefficient the federal procurement process was. The solar heat technologies program was implemented to a very large degree through small contracts to individuals, small businesses, not-for-profit organizations, universities, and some large companies. During the first five years of the program, there were hundreds of contracts, most of them with values of less than $100,000. From our perspective as program managers, and more recently as contractors, we know how frustrating the procurement process is to all parties. Procedures intended to protect the taxpayers' interests and to assure the government gets full value for its dollars often have just the opposite effect. There are three major problems with the federal procurement system as applied by DOE in the solar energy program: (1) it is too slow, (2) it is too cumbersome for small contracts, and (3) contracts are administered by offices that have no responsibility for, and little interest in, the outcome of the contract or the program it supports.

All government agencies operate on a one-year budget cycle. In theory, each agency receives its budget allocation on or before the beginning of the fiscal year (now October 1), and must have spent or committed those funds by September 30 of the following year. In fact, it is often several months into the fiscal year before agencies or offices within agencies know their budgets and receive the authority to spend them. This problem is compounded by the fact that procurement times (the time between procurement announcement and contract award) are rarely less than 6 months and frequently 12 to 18 months. This situation puts managers in a totally untenable position. They must plan programs to achieve certain objectives within the budget year, for which they are accountable, but they are unable to implement them by the expected time. As a result, both managers and contractors are always under unreasonable and unnecessary time pressures. Managers spend an excessive amount of their time trying to "beat the system" in placing crucial contracts. Bidders are often forced to exaggerate their capabilities, especially schedules, in order to win a contract; as contractors, they may produce less than their best work in order to meet an impossible schedule. Although it makes no sense for the procurement process to take longer than the execution time for an R&D project, this is often the case. R&D is not a commodity that can be shipped whenever the order is received.

The second major problem is the indiscriminate application of contract regulations designed for major government contracts to all DOE contracts, large and small. Small businesses, or individuals, are often asked to comply with accounting practices designed to keep a large defense contractor honest. Conforming to these rules is often an extremely costly burden on small businesses and an unacceptable intrusion on larger ones. For example, the rules on accounting for cost reimbursable contracts are only intended for contracts over $500,000, but DOE field offices applied them to all cost-reimbursable contracts. This is simply an example of applying the strictest rule available to all situations in order to avoid criticism for being too loose with the taxpayers' money. No one ever calculated the additional cost to taxpayers of added contractor and government overhead, sour contractor relations, and lost resources caused by following these rules.

The first two problems are exacerbated by the third. Procurement officials in government are physically, administratively, and mentally removed from the technical program managers. Even when procurement officers have the authority to apply the rules to the benefit of the programs, or to use more efficient contracts, few will risk potential criticism for the sake of a project or program that is of no direct concern to them. The take-no-risk syndrome, present in all branches of government, reaches its pinnacle in procurement offices.

Note

1. Today the NASA-industry partnership is under attack, and the lessons learned from more than seventy-five years of cooperation are in danger of being lost. For example, if a U.S. aircraft designer wants special wind tunnel tests performed on its latest airframes, the best facilities are now in Europe, not, as previously, at a NACA/NASA laboratory. These labs may soon go the way of the Wright Flyer.

Appendix: Outline of the Solar Heat Technologies Series

Volume 2: Solar Resources, edited by Roland L. Hulstrom

1. **Introduction**
   Roland L. Hulstrom
2. **Insolation Data Bases and Resources in the United States**
   Raymond J. Bahm
3. **Insolation Models and Algorithms**
   Charles M. Randall and Richard Bird
4. **Solar Radiation Monitoring Networks**
   Kirby Hanson and Thomas Stoffel
5. **Solar Radiation Instrumentation**
   Gene Zerlaut
6. **Spectral Terrestrial Solar Radiation**
   Richard Bird
7. **Insolation Forecasting**
   John Jensenius
8. **Illuminance Models and Resources in the United States**
   Claude Robbins

Volume 3: Economic Analysis of Solar Thermal Energy Systems, edited by Ronald E. West and Frank Kreith

1. **Introduction**
   Ronald E. West
2. **Economic Methods**
   Rosalie T. Ruegg and Walter Short
3. **Economic Models**
   G. Thomas Sav
4. **Assessing Market Potential**
   Gerald E. Bennington
5. **Analyzing the Effect of Economic Policy on Solar Markets**
   Peter C. Spewak
6. **End-Use Matching and Applications Analysis Methodologies**
   Kenneth C. Brown
7. **Net Energy Considerations**
   Robert A. Herendeen
8. **Cost Requirements for Active Solar Heating and Cooling**
   Mashuri L. Warren
9. **Cost Requirements for Passive Solar Heating and Cooling**
   Charles R. Hauer

10	Cost Requirements for Solar Thermal Electric and Industrial Process Heat Ronald Edelstein
11	Historical Cost Review Charles E. Hansen and Wesley L. Tennant

Volume 4: Fundamentals of Building Energy Dynamics, edited by Bruce D. Hunn

1	Introduction: Energy Use in Buildings Bruce D. Hunn
2	Patterns of Energy Use in Buildings Arthur H. Rosenfeld, Mark D. Levine, Evan Mills, and Bruce D. Hunn
3	Characterization of Energy Processes in Buildings Robert D. Busch
4	Methods of Energy Analysis Robert D. Busch
5	Energy Conservation and Management Strategies P. Richard Rittelmann

Volume 5: Solar Collectors, Energy Storage, and Materials, edited by Francis de Winter

I SOLAR COLLECTORS

1	Overview Francis de Winter
2	Collector Concepts and Designs Ari Rabl
3	Optical Theory and Modeling of Solar Collectors Ari Rabl
4	Thermal Theory and Modeling of Solar Collectors Noam Lior
5	Testing and Evaluation of Stationary Collectors Robert D. Dikkers
6	Testing and Evaluation of Tracking Collectors David W. Kearney
7	Optical Research and Development Roland Winston
8	Collector Thermal Research and Development Charles A. Bankston

Outline of the Solar Heat Technologies Series

9   **Collector Engineering Research and Development**
Charles F. Kutscher

10   **Solar Pond Research and Development**
John R. Hull and Carl E. Nielsen

11   **Cost Issues and Opportunities**
John A. Clark

12   **Reliability and Durability of Solar Collectors**
William Freeborne

13   **Environmental Degradation of Low-Cost Solar Collectors: Research Issues and Opportunities**
Fred Loxsom and Eugene Clark

II   **ENERGY STORAGE FOR SOLAR SYSTEMS**

14   **Overview**
C. J. Swet

15   **Storage Concepts and Design**
C. J. Swet

16   **Analytical and Numerical Modeling of Thermal Energy Storage**
John R. Hull

17   **Testing and Evaluation of Thermal Energy Storage Systems**
Robert D. Dikkers

18   **Storage Research and Development**
Allan I. Michaels and C. J. Swet

19   **Issues and Opportunities**
C. J. Swet

III   **MATERIALS FOR SOLAR TECHNOLOGIES**

20   **Overview**
Stanley W. Moore

21   **Materials for Solar Collector Concepts and Designs**
Richard Silberglitt and Hien K. Le

22   **Theory and Modeling of Solar Materials**
Carl M. Lampert

23   **Testing and Evaluation of Solar Materials**
Hien K. Le and Richard Silberglitt

24   **Exposure Testing and Evaluation of Performance Degradation**
Thomas E. Anderson

25   **Solar Materials Research and Development**
Richard Silberglitt and Hien K. Le

26   **Solar Materials Issues and Opportunities**
C. J. Swet

Volume 6: Active Solar Systems, edited by George Löf

I	DESIGN, ANALYSIS, AND CONTROL METHODS
1	Overview of Modeling and Simulation John A. Duffie
2	Detailed Simulation Methods and Validation William A. Beckman
3	Design Methods for Active Solar Systems S. A. Klein
4	Controls in Active Solar Energy Systems C. Byron Winn
5	Optimum System Design Techniques Jeff Morehouse
II	SOLAR WATER HEATING
6	Solar Water Heating—Introduction and Summary George Löf
7	Hot Water Systems: System Concepts and Design Elmer Streed
8	The Performance of Residential Solar Hot Water Systems in the Field William Stoney
9	Mechanical Performance and Reliability R. M. Wolosewicz
10	Costs of Commercial Solar Hot Water Systems Thomas King and Jeff Shingleton
III	SOLAR SPACE HEATING
11	Space Heating Systems—Introduction and Summary George Löf
12	Space Heating: System Concepts and Design S. Karaki
13	System Performance Charles Smith
14	Performance of Residential and Commercial Systems S. M. Embrey
15	Costs of Solar Space Heating Systems Thomas King and Jeff Shingleton
16	Community Scale Systems Seasonal Thermal Energy Storage and Central Solar Heating Plants with Seasonal Storage Dwayne S. Breger

IV	SOLAR COOLING
17	Research and New Concepts Noam Lior
18	Solar Cooling—Introduction and Summary Michael Wahlig
19	Mechanical Systems and Components Henry M. Curran
20	Absorption Systems and Components Michael Wahlig
21	Desiccant Systems and Components John Mitchell
22	Comparative Performance and Costs of Solar Cooling Systems Mashuri Warren

Volume 7: Passive Solar Buildings, edited by J. Douglas Balcomb

1	Introduction J. Douglas Balcomb
2	Building Solar Gain Modeling Patrick J. Burns
3	Simulation Analysis Philip W. B. Niles
4	Simplified Methods G. F. Jones and William O. Wray
5	Materials and Components Timothy E. Johnson
6	Analytical Results for Specific Systems Robert W. Jones
7	Test Modules Fuller Moore
8	Building Integration Michael J. Holtz
9	Performance Monitoring and Results Donald J. Frey
10	Design Tools John S. Reynolds

Volume 8: Passive Cooling, edited by Jeffrey Cook

1	Introduction Jeffrey Cook

2   Ventilative Cooling
    Subrato Chandra
3   Evaporative Cooling
    John I. Yellott
4   Radiative Cooling
    Marlo Martin
5   Earth Coupling
    Kenneth Labs
6   Passive Cooling Systems
    Gene Clark
7   The State of Passive Cooling Research
    Jeffrey Cook

Volume 9: Solar Building Architecture, edited by Bruce Anderson

1   Introduction
    Bruce Anderson
2   Site, Community, and Urban Planning
    Layne Ridley
3   Building Envelopes
    Donald Prowler and Douglas Kelbaugh
4   Thermal Energy Storage in Building Interiors
    Bion D. Howard and Harrison Fraker
5   Thermal Energy Distribution in Building Interiors
    Gregory Franta
6   Architectural Integration: Residential and Light Commercial Buildings
    Edward Mazria
7   Nonresidential Buildings
    Harry T. Gordon, John K. Holton, N. Scott Jones, Justin Estoque, Donald L. Anderson, and William J. Fisher

Volume 10: Implementation of Solar Thermal Technology, edited by Ronal Larson and Ronald E. West

I   SOLAR THERMAL TECHNOLOGIES IMPLEMENTATION

1   Introduction
    Ronal W. Larson and Ronald E. West
2   The Role of Congress
    J. Glen Moore
3   Market Development
    Carlo Laporta

Outline of the Solar Heat Technologies Series

II		**SOLAR THERMAL PROGRAMMATIC PERSPECTIVES**
	4	**Active Heating and Cooling** William Scholten
	5	**Passive Technologies** Mary Margaret Jenior and Robert T. Lorand
	6	**Passive Commercial Buildings Activities** Robert G. Shibley
	7	**Industrial Process Heat** David W. Kearney
	8	**High-Temperature Technologies** J. C. Grosskreutz
III		**SOLAR THERMAL DEMONSTRATIONS AND CONSTRUCTION**
	9	**Residential Buildings** Murrey D. Goldberg
	10	**Commercial Buildings** Myron L. Myers
	11	**Federal Buildings** Oscar Hillig
	12	**Agricultural Demonstration Programs** Robert G. Yeck and Marvin D. Hall
	13	**Military Demonstration Programs** William A. Tolbert
IV		**SOLAR THERMAL QUALITY**
	14	**Testing, Standards, and Certification** Gene A. Zerlaut
	15	**Consumer Assurance** Roberta W. Walsh
V		**SOLAR THERMAL INFORMATION**
	16	**Consumer Information** Rebecca Vories
	17	**Public Information** Kenneth Bordner and Gerald Mara
	18	**Technical Information** Paul Notari
	19	**Training and Education** Kevin O'Connor
	20	**Regional Solar Energy Centers** Donald E. Anderson

VI SOLAR THERMAL TECHNOLOGY TRANSFER

21 Liaison with Industry
Daniel Halacy

22 Solar Energy Research Institute
Barry L. Butler

23 Los Alamos National Laboratory
J. Douglas Balcomb and W. Henry Lambright

24 Argonne National Laboratory
William W. Schertz

VII SOLAR THERMAL INCENTIVES

25 Tax Credits
Daniel Rich and J. David Roessner

26 Financing
Steven Ferrey

27 Grants
Seymour Warkov and T. P. Schwartz

VIII SOLAR THERMAL ORGANIZATIONAL SUPPORT

28 International Activities
Murrey D. Goldberg

29 State and Local Activities
Peggy Wrenn and Michael DeAngelis

30 Public Utilities
Stephen L. Feldman and Patricia Weis Taylor

31 Legal, Environmental, and Labor Issues
Alan S. Miller

Contributors

Charles A. Bankston
Charles A. Bankston, the editor-in-chief of this series, has been active in solar energy research and development for more than twenty years. Bankston was professionally associated with the Los Alamos National Laboratory, where he conducted directed research in heat transfer, nuclear propulsion, advanced drilling, geothermal energy, and solar energy from 1958 to 1982. In 1982 Bankston founded CBY Associates, an energy consulting firm of which he is the president. While at Los Alamos, Bankston was responsible for technical direction of the U.S. DOE program for solar collector and materials research—a program that involved more than one hundred research contracts. His personal research in solar energy applications includes studies of large collector arrays, concentrating collectors, seasonal thermal energy storage, and systems analysis and optimization. Bankston has published numerous papers and articles, and was an associate editor of *Solar Energy Engineering*. He received his D.Sc. degree in mechanical engineering from the University of New Mexico.

Donald A. Beattie
Don Beattie has a wide-ranging background in R&D management in both the private and public sectors spanning thirty-eight years. Coming

from an oil exploration background, he joined NASA in 1963 at the beginning of the Apollo Program and eventually served as program manager for Apollo lunar surface experiments. At the conclusion of the Apollo program he joined the National Science Foundation and began a ten-year career managing federal energy RD&D programs, including the initial build-up of the NSF, ERDA, and DOE solar energy programs, overseeing their growth from $4 million in FY 1973 to $500 million in FY 1979. Solar heat technologies programs, through these years, received a majority of the solar program funds. After receiving his A.B. degree from Columbia University, he served as a naval aviator, and upon completion of his naval service received an M.S. degree from the Colorado School of Mines.

Frederick H. Morse
Frederick Morse received his undergraduate degree in mechanical engineering from Rensselaer Polytechnic Institute, a master's in nuclear engineering from Massachusetts Institute of Technology, and a doctorate in mechanical engineering from Stanford University. After holding a teaching position at the University of Maryland, Dr. Morse joined the U.S. Department of Energy, where he had a significant role in all management aspects of the U.S. solar energy program since its inception in 1972.

At DOE, he held three senior management positions: Director of the Office of Solar Heat Technologies, Director of the Office of Solar Applications, and Chief of Solar Heating and Cooling. He was responsible for the development, implementation, and management of a highly diversified major government solar energy research, development, and commercialization program with annual budgets of up to $250 million. Dr. Morse had major responsibility for the active, passive, solar thermal electric, and photovoltaic technologies. He managed a staff of forty-five professionals and the work of several national laboratories, DOE field offices, and regional solar energy centers; he worked closely with industry, utilities, and other elements of the private sector to develop long-range plans for technology development that reflected industry needs and viewpoints.

Since 1989, Dr. Morse has been president of Morse Associates, Inc., an energy and environmental consulting firm serving domestic and international, government and private business clients.

Index

AAI Corporation, 36
Accelerated Program, 33
Accelerated R&D scenario, 41–43
Active Solar Systems, 207
Active systems, 213–214
 domestic hot water, 214–217
 space cooling and heating, 219–220
 space heating, 217–219
 swimming pool heating, 214
Adams, Martin, 91–92
Advanced Research Project Agency (ARPA), 32
Agnew, Harold, 123
Air-Conditioning and Refrigeration Institute (ARI), 134
Alad'yev, Ivan T., 50
Albuquerque Operations Office (AOO), 12
Alliance to Save Energy (ASE), 232
Alm, Alvin L., 98, 105
Alternatives and Recommendations Related to the Mission, Role, and Management of the Solar Energy Research Institute, 81
American Institute of Architects (AIA), 116
American Society of Heating, Refrigerating, and Air-Conditioning Engineers (ASHRAE), 117, 194
American Solar Energy Society Round Table on Societal Costs, 241
Anderson, Bruce, 206
Application of Solar Technology to Today's Energy Needs, 104, 127
Aquifer storage, 205
Architecture and planning, 220
Argonne National Laboratory (ANL), 12, 121, 148
Association for Applied Solar Energy, 25
Atomic Energy Commission (AEC), 30, 95
Automobile fuel efficiency, 248

Balcomb, J. Douglas
 on linking of technologies, 70
 on passive solar technology, 207
 resignation letter by, 123
 at SERI, 147
Baldridge, Malcolm, 153
Barriers and Incentives Branch, 89
Battelle Laboratories, 88
Baum, V. A., 50
Bayh-Dole Act (PL 96-517), 165
Beattie, Donald A., 101
 at DOE, 109–110
 on IPTASE panel, 32
 and RSECs, 129
Beggs, James, 148

Bezdek, Roger H., 89
Bicentennial Solar and Energy Conservation Exhibit, 88
Block, John, 153
Boer, Karl, 26
Bowden, Donald, 55
Brookhaven National Laboratory (BNL), 12, 121, 148
Budgets, 6–9, 93, 95–97
 in Bush administration, 178
 in Clinton administration, 180
 multiyear funding, 254
 in Reagan administration, 170–176
Building Energy Performance Standards (BEPS), 146
Buildings
 energy dynamics of, 207–209
 solar technologies for, 206–207
Bumpers, Dale, 174
Bureau of the Budget (BOB), 95
Bureau of Land Management (BLM), 15
Bush administration, 177–178
Business as Usual R&D scenario, 41–43
Busterud, John, 139–140
Butler, Barry, 147
Butt, Sheldon, 118, 131–133

Cabot Fund, 24
Carter, Hugh A., Jr., 99
Carter administration, 109–110
 cabinet changes in, 118–120
 commercialization and market development in, 113–116
 and DOE establishment, 97–108
 ERDA studies in, 90–93
 field support structure in, 120–131
 legislation during, 135–136
 Office of Conservation and Solar Applications in, 111–113
 results from, 246–247
 and SEIA, 131–135
 SFBP program in, 116–117
 solar dedication speech, 117–118
 Walden appointment, 110–111
"Catalog on Solar Energy Heating and Cooling Products," 87
Central receiver collectors, 197, 224–225
Certification programs, 133–134, 194–195
Chemical storage, 206
China, 52
Clean Coal Program, 249–251
Climate Change Action Plan, 179
Clinton administration, 178–180

Codes
 evaluation of, 229–230
 by SEIA, 133–135
Cohen, Larry, 22
Collection and dissemination of information, 76–77
Collectors
 advanced development in, 197–200
 concepts of, 191–192
 cost reduction R&D for, 200–203
 optical and thermal analysis of, 192–194
 performance and quality control for, 195–196
 tests and standards for, 194–195
Collectors, Energy Storage, and Materials, 221
Collins, Pat, 158, 162
Commercialization
 agencies responsible for, 73–75
 under Carter administration, 113–116
 by FEA, 65
 government role in, 251–252
 of solar water heating, 103–104
 support structure for, 10–14
"Commercialization Plan for Solar Water Heating," 103–104
Commission on the Challenges of Modern Society (CCMS), 48–49
Committee on Energy R&D Goals, 27
Committee on Public Engineering Policy (COPEP), 29
Committee on Renewable Energy Commerce and Trade (CORECT), 235
Compound parabolic concentrator (CPC) collectors, 192–193, 198–199
Computer ray trace techniques, 192–193
Conference on New Energy Sources, 25
Congress
 budgets, 6–9, 93, 95–97, 170–176, 178, 180, 254
 legislation, 2–4, 45–47, 135–136, 166–168
Conservation and Renewable Energy Office, 178
Conservation efforts, 248
Construction programs, 227–229
Controlled Thermonuclear Research program, 56
Cook, Jeffrey, 207, 213
Cooling systems, 212–213, 219–220
Corporate Average Fuel Economy (CAFE) standards, 248
Council on Environmental Quality (CEQ), 104
Cranston, Alan, 174

Crude Oil Windfall Profit Tax Act (PL 96-223), 136
Data bases, 125, 150, 181–183
 ERSATZ stations, 182
 Solar Energy Information Data Bank, 125, 150
Davis, J. Michael, 177–178
DeConcini, Dennis, 174
DeGeorge, Frank, 149
Demand-side management (DSM), 159
Demonstration programs, 227–229
 agencies responsible for, 72–75
 by ERDA, 62–67
 NSF plans for, 46
 phaseout of, 114
Denmark
 in CCMS, 48
 energy taxes in, 244
Department of Agriculture (USDA), 32
Department of Commerce (DOC), 32, 141
Department of Defense (DOD)
 collection and dissemination of information, 76–77
 commercial demonstration programs, 73–75
 research and development, 75–76
 residential demonstration programs, 72–73
Department of Energy (DOE), 7, 10
 attempt to close, 141–142
 under Edwards, 140–141
 establishment of, 97–108
 field offices (FO), 12
 Herrington appointment, 162
 Hodel appointment, 158
 O'Leary appointment, 179
 organization of, 10–13
Department of Energy Act (PL 94-238), 136
Department of Energy Organization Act (PL 95-91), 99–100, 136
Department of Housing and Urban Development (HUD), 30
 collection and dissemination of information, 76–77
 commercial demonstration programs, 73–75
 research and development, 75–76
 residential demonstration programs, 72–73
Department of Interior (DOI), 30
Deutch, John, 123–124
de Winter, Francis, 187
Dingell, John, 172
Directorate for Research Applications, 26
Disincentives, 242–243

Index   271

Doctor, Ronald, 107
DOE-OMB Budget Cases document, 145
Domestic hot water (DHW) systems, 204, 214–217
Domestic Policy Review (DPR), 104–107
"Domestic Policy Review of Solar Energy," 90, 105–106
Duffie, John, 26
Duncan, Charles, 118

Earth in the Balance, 179
Economic Analysis of Solar Thermal Energy Systems, 185
Economic Analysis of Solar Water and Space Heating, An, 69
Economics, 183, 185–186
Education and training programs, 135, 230–231
Edwards, James
  appointment of, 140–141
  budget cuts under, 144
  and Solar Bank, 153
Eggers, Alfred J., Jr., 27, 33–34, 47, 57
Eizenstat, Stuart, 104
Energy Conservation and Production Act (ECPA) (PL 94-385), 65, 71, 97
Energy Efficiency Ratings (EER), 134
Energy Information Administration (EIA), 133
Energy Policy Act (PL 102-486), 177
Energy Policy Council, 101
Energy Policy Project, 53–54
Energy Reorganization Act (PL 93-438), 45, 55
Energy Research Advisory Board Solar Panel, 163–164
Energy Research and Development Administration (ERDA), 31, 40
  budget preparation for, 95–96
  collection and dissemination of information, 76–77
  commercial demonstration programs, 73–75
  and DOE, 99
  ERDA-49 report, 62–67
  ERDA 76-1 report, 85–87
  establishment of, 55–59
  IERS study, 92–93
  MOPPS study, 90–92
  and national laboratories, 85
  national plan by, 59–62, 67–78
  R&D in, 75–76
  RD&D in, 246
  residential demonstration programs, 72–73
  technology assessments by, 186

Energy Research and Technology Administration (ERTA), 141
Energy Resources Council, 55
Energy Security Act (PL 96-294), 152–153
Energy Security report, 165
Energy storage, 203–206
Energy Tax Act (PL 95-618), 132, 136
Environmental issues
  in Project Independence, 43
  vs. security, 176
Environmental Policy Center, 118
Environmental Protection Agency (EPA), 32
ERDA-23 report, 68
ERDA-23A report, 70, 89
ERDA-48 report, 60–62
ERDA-49 report, 62–67
ERDA-75 report, 87
ERDA 76-1 report, 85–87
ERDA 76-144 report, 186
ERDA Source Evaluation and Selection Handbook, 81
ERSATZ stations, 182, 184
Ervin, Christine, 180
Evacuated tube collectors, 197–198
Evans, Dan, 173

Farber, Eric, 26
Fauquier High School, 36–38
Federal buildings program, 71, 116–117, 228
Federal Council on Science and Technology (FCST), 27
Federal Energy Administration (FEA), 40–41
  collection and dissemination of information, 76–77
  commercial demonstration programs, 73–75
  and ERDA, 65–66
  research and development, 75–76
  residential demonstration programs, 72–73
Federal Energy Administration Act (PL 93-275), 45
Federal Energy Management Program (FEMP), 71
Federal Energy Office (FEO), 40–41
Federal Nonnuclear Energy Research and Development Act (PL 93-577), 45, 59, 85
Fields, Raymond H., 48, 81
Field support structure, 120–121
  national laboratories, 121–124
  in Reagan administration, 147–149
  RSECs, 128–131
  SERI, 124–128

Fitzpatrick, Donna, 162–165, 167, 174–175
Fletcher, James E., 30–31
Ford administration
  barriers and incentives in, 88–90
  demonstration programs, 62–67
  ERDA under, 55–59, 84–88
  national energy plan, 59–62
  national solar heating and cooling program, 67–78
  SERI under, 78–83
Ford Foundation, 52–54
Fowler, Wyche, 173–174
France
  agreements with, 51
  in CCMS, 48
Freeman, S. David, 53
Fri, Robert W., 82
  and ERDA, 99
  and IERS study, 92
  and MOPPS study, 91
Fulmer, Bert, 154
Fundamentals of Building Energy Dynamics, 206–208

Garwin, Richard L., 79
General Accounting Office (GAO), 130–131
General Agreement on Tariffs and Trade (GATT), 250
General Electric, 35, 37, 52
General Services Administration (GSA)
  collection and dissemination of information, 76–77
  commercial demonstration programs, 73–75
  research and development, 75–76
  residential demonstration programs, 72–73
General Technology for the Utilization of Solar Energy, 49
Geothermal Energy Research, Development and Demonstration Act (PL 93-410), 45
Germany, energy taxes in, 244
Glazer, Peter, 198
Goldwater, Barry M., Jr., 62
Gore, Al, 178–179
Gouraud, Jackson S.
  appointment of, 103
  and commercialization, 112–113
  and RSECs, 130
  and tax credits, 132
Government, 237–239
  administration changes in, 252–254
  energy policies by, 239–244
  RD&D role of, 245–252
Government National Mortgage Association (GNMA), 154

Grant Review Board, 30
Grants from RANN, 30
Green, Richard J., 49
Grosskreutz, Charles, 126
Grover Cleveland Junior High School, 36–37
Growth scenarios, 52–54

Handler, Philip, 79
Haskell, Floyd K., 125
Hatfield, Mark, 173
Hayes, Denis, 126–127, 130, 149
Heating, ventilation, and air-conditioning (HVAC), 208
Heliotek, 52
Heritage Foundation, 16, 139, 155
Herrington, John S., 162, 165, 168–169
Herwig, Lloyd O., 30
Hirsch, Robert S., 84, 87, 96, 99
Hodel, Donald, 158, 162, 171
Holifield, Chet, 95
Holifield Tables, 95–96
Holmes, Jay E., 20
Honeywell, 37, 52
Horowitz, Harold, 30
Hottel, Hoyt C., 24, 192
Hot water systems, 214–217
House Committee on Oversight and Investigations, 112
House Committee on Science and Astronautics, 28
Housing and Community Development Act (PL 93-383), 45, 63
Housing and Urban Development (HUD), 30
  collection and dissemination of information, 76–77
  commercial demonstration programs, 73–75
  research and development, 75–76
  residential demonstration programs, 72–73
Hubbard, Harold, 149–150
Hulstrom, Roland L., 182
Humphrey, Hubert, 82
Hunn, Bruce D., 206

Ideas for Tomorrow, Choices for Today, 160
Implementation of Solar Thermal Technology, 221, 227
Ince, James, 131
Incentives, 89–90, 242–244
  for domestic hot water systems, 215
  evaluation of, 232–235
  in Ford administration, 89
  in Project Independence, 43

in Reagan administration, 168–170
SEIA support for, 131–133
for space heating systems, 217
Industrial process heat, 88, 225–226
Industrial Process Heat Program, 88
Inexhaustible Energy Resources Study (IERS), 91–93
Information programs, 230–231
agencies responsible for, 76–77
by RSECs, 150–151
Instability from administration changes, 252–254
Institute of Gas Technology, 67
Integrated collector and storage (ICS) systems, 216
Integrated CPC evaluated collectors (ICPCs), 198
Integrated resource planning (IRP), 178
Interagency Panel for the Terrestrial Applications of Solar Energy (IPTASE), 31–32
Interdisciplinary Research Relevant to Problems of Our Society (IRRPOS), 26, 29
International Energy Agency (IEA), 49, 235–236
International energy cooperation, 48, 51–52, 235–236
CCMS, 48–49
SOLERAS program, 51–52
with USSR, 49–51
International Solar Energy Society, 25
International Solar Industry Expo 75, 61
Interstate Solar Coordinating Committee (ISCC), 134
InterTechnology Corporation, 37, 88

Japan, research program by, 52
Jet Propulsion Laboratory (JPL), 31, 122, 148
Job Corps, 135
Johnson, Harry R., 91
Johnson, R. Tenney, 59
Johnson Space Flight Center, 31
Joint Commission on Economic Cooperation, 51
Joint Commission on Scientific and Technical Cooperation, 49
Joint Committee on Atomic Energy (JCAE), 95

Kane, James S., 92
Karpe, Robert, 154
Kelly, Henry, 127–128

Kerr, Donald, 123
Kreith, Frank, 125, 185

Labeling programs, 133–134
Langley Research Center, 31
Large-scale solar school projects, 36–39
Larson, Ronal, 226–227
Lawrence Berkeley Laboratory (LBL), 121, 148
Lawrence Livermore Laboratory (LLL), 121–122, 148
Le Gassie, Roger W. A., 60, 91
Legislation, 2–4, 45–47
during Carter administration, 135–136
during Reagan administration, 166–168
Lewis Research Center, 31
Lloyd, Marilyn, 167
Loan guarantee programs, 63
Löf, George, 26, 207
research house by, 217
on solar priorities, 106–107
Lorsch, Harold, 26
Los Alamos National Laboratory (LANL), 121
Los Alamos Scientific Laboratory (LASL), 121, 123, 148
Lovelace, Alan M., 102
LS3 collectors, 222–223
LS4 collectors, 223
Luz International, 198, 221–225, 234–235

Magnetohydrodynamics, 30
Management Operations Study Task Force, 129
Mandate for Leadership, 139, 155
Mark, Hans, 148
Market development under Carter administration, 113–116
Market-Oriented Program Planning Study (MOPPS), 90–92
Marshall Space Flight Center (MSFC), 31, 122
Materials
advances in, 187, 190–191
research in, 251
Materials Policy Commission, 25, 27
Matsunaga, Spark, 174
McClure, James, 141–142
McCormack, Mike, 47, 62
McGee, Robert P., 79
McKinney, Stewart B., 152
Mid-America Solar Energy Complex (MASEC), 150–151

Midwest Regional Solar Energy Planning Venture, 128
Midwest Research Institute (MRI), 82, 124
Miller, Bennett, 92
Minimum Viable Program, 33
MITRE Corporation, 79–80
Molten-salt storage, 204
Morse, Frederick H., 1, 11
 documentation objectives of, 17–20
 in international pilot study, 48
 on RANN staff, 30
Multiyear funding, 254
Munson, Richard, 107
Myers, Dale E.
 at DOE, 100–102
 and SERI and RSEC roles, 125–126

National Academy of Engineering (NAE), 28–29, 79
National Academy of Sciences (NAS), 52, 79
National Advisory Council on Aeronautics (NACA), 250
National Aeronautics and Space Administration (NASA), 30
 collection and dissemination of information, 76–77
 commercial demonstration programs, 73–75
 Memorandum of Understanding (MOU), 31, 102
 and NSF, 31
 research and development, 75–76
 residential demonstration programs, 72–73
 termination of energy programs, 148
National Association of Home Builders, 151
National Bureau of Standards (NBS), 30
 collection and dissemination of information, 76–77
 and collectors, 194
 commercial demonstration programs, 73–75
 research and development, 75–76
 residential demonstration programs, 72–73
National Center for Scientific Research (CNRS), 51
National Energy Act, 112
National Energy Conservation Policy Act (PL 95-619), 136, 155–156
National Energy Plan, 98
National Energy Policy Act, 167
National Energy Policy Plan, The, 158–159
National Energy Research Act, 132
National Energy Strategy report, 177

National Institute of Standards and Technology (NIST), 194
National laboratories
 and ERDA, 85
 roles of, 121–124
National Oceanic and Atmospheric Administration (NOAA), 182
National Plan for Accelerated Commercialization of Solar Energy, 112
National Plan for Solar Heating and Cooling, 68
National Program Plan for Research and Development in Solar Heating and Cooling, 186–189
National Renewable Energy Laboratory (NREL), 14, 78, 178
National Science Foundation (NSF)
 collection and dissemination of information, 76–77
 commercial demonstration programs, 73–75
 demonstrations planned by, 46
 early support for, 26–29
 and ERDA, 55–59
 as lead energy agency, 29–32
 and Project Independence, 41
 research and development, 75–76
 residential demonstration programs, 72–73
National Science Foundation Act (PL 81-507), 58
National Science Foundation Authorization Act (PL 94-413), 54–55
National Solar Energy Program, 89
National Solar Radiation Data Base for the United States, 183
National Weather Service (NWS), 182
Nation's Energy Future, The, 33–34, 40, 42
Neal, Stephen L., 152
Nixon administration
 energy studies, 27
 expansion of energy programs in, 33–35
 Ford Foundation project during, 52–54
 international energy cooperation under, 48–52
 national solar energy program, 54–55
 NSF as lead energy agency in, 29–32
 NSF support under, 26–29
 and Project Independence, 39–44
 solar school projects, 36–39
Noland, Michael, 124
North Atlantic Treaty Organization (NATO), 48
Northeast Solar Energy Center (NESEC), 151–152

Index 275

Northern Energy Corporation, 128
Northview Junior High School, 36–38
Nuclear Regulatory Commission (NRC), 55

Ocean thermal energy conversion (OTEC),
Office of Building Technologies, 178
Office of Coal Research (OCR), 30
Office of Commercialization, 113
Office of Conservation and Solar
 Applications, 111–113, 126
Office of Energy Technology, 126
Office of Management and Budget (OMB),
 6, 95, 142–144, 160–161
Office of Saline Water, 23
Office of Science and Technology (OST), 27
Office of Solar Applications, 113–114
Office of Solar Heat Technologies, 143
Office of Technology Assessment (OTA),
 104
Office of Utility Technologies, 178
OG-100 standard, 194
OG-300 standard, 195, 216
Oil embargo, 23
O'Leary, Hazel, 179–180
O'Leary, John, 101, 119
O'Neill, Tip, 82–83
Orbiting mirrors, 200
Organization of Petroleum Exporting
 Countries (OPEC) oil embargo, 23, 33,
 238
Organizational support programs, 235–236
Orphan systems, 170
Ottinger, Richard, 146
Oversight and Investigations committee, 112

Parabolic dish collectors, 197
Parabolic trough collectors, 197–198, 222–
 223
Parasitic power losses, 196
Passive Cooling, 207
Passive cooling systems, 207, 212–213
Passive heating systems, 209–212
Passive Solar Buildings, 207
Pastore, Richard, 34
Patents
 in ERDA, 59, 64
 in Reagan administration, 165
People's Republic of China, 52
Performance of collectors, 195–196
Phase-change storage, 203, 205–206
Photovoltaic (PV) technologies, 119, 148
Physical Research program, 56
Pierce, Samuel, 153
Plastic films, 201
Power technologies, 221–225, 234–235

*Priorities for Research Applicable to
 National Needs,* 29
Procurement process, 255–256
Program Opportunity Announcements
 (POAs), 59
Program Opportunity Notices (PONs), 59
Program Research and Development
 Announcement (PRDA), 186
Program solicitations from RANN, 30
Project Independence, 33–35, 39–40
Project Independence Blueprint, 40–44
Project Independence Solar Energy Task
 Force, 39, 42–43
Project Sunshine, 52
Proof of Concept Experiments (POCEs), 35,
 163
Public information programs, 230–231
Public Utilities Regulatory Policy Act
 (PURPA) (PL 95-617), 90, 234

Quality assurance programs, 229–230
Quality control for collectors, 195–196

Rappaport, Paul, 83, 124–125, 127
Ratchford, Thomas P., 57
Rating, certification, and labeling (RCL)
 programs, 133–134
Ratings for collectors, 194–195
Ray, Dixie Lee, 33–34
Ray trace techniques, 192–193
Reagan administration, 16
 budgets in, 170–176
 Collins appointment, 158
 DOE under, 141–142
 Edwards appointment, 140–141
 energy and security under, 165–166
 field support in, 147–149
 Fitzpatrick appointment, 162–165
 Herrington appointment, 162
 Hodel appointment, 158
 Hubbard appointment, 149–150
 legislation during, 166–168
 OMB in, 142–144, 160–161
 patent policies in, 165
 RCS in, 155–156
 reductions in force in, 146–147
 results from, 247–248
 RSECs in, 148–152
 SERI in, 147–150
 Solar Bank in, 152–155
 status at end of first term, 159–160
 tax credits in, 168–170
 transition team, 139–140
 Tribble appointment, 145–146
 Tribble departure, 156–158

Reductions in force (RIF) in Reagan administration, 17, 146–147
Regan, Donald, 153
Regional Solar Energy Centers (RSECs), 83
  creation of, 128–130
  management of, 130–131
  in Reagan administration, 148–152
Renewable Energy and Conservation Transition Act, 168
Renewable Energy and Energy Conservation Commercialization and Development Act, 167
Request for Proposals (RFP), 55, 82
Research and development (R&D)
  agencies responsible for, 75–76
  for collectors, 200–203
  scenarios for, 41–43
Research Applied to National Needs (RANN), 26–27, 29–32
  activities at, 35
  large-scale solar school projects by, 36–39
  support for, 85
Research, development, and demonstration (RD&D), 1
  government role in, 245–252
  support structure for, 10–14
Residential Conservation Service (RCS), 155–156
Residential demonstration programs, agencies responsible for, 72–73
Results of solar programs, 181
  active systems, 213–220
  collectors, 191–203
  demonstration and construction programs, 227–229
  economics, 183, 185–186
  energy dynamics of buildings, 207–209
  energy storage, 203–206
  implementation, 226–236
  incentive programs, 232–235
  in industry, 225–226
  information programs, 230–231
  materials, 187, 190–191
  organizational support programs, 235–236
  passive cooling systems, 212–213
  passive heating systems, 209–212
  power systems, 221–225
  quality assurance programs, 229–230
  solar architecture and planning, 220
  solar resource analysis, 181–184
  solar technologies for buildings, 206–207
  technology, 186–189
  technology transfer programs, 231
Runnels, Harold, 90

Salgado, Joseph, 163
Sandia National Laboratory Albuquerque (SNLA), 12, 51
Sandia National Laboratory Livermore (SNLL), 12, 121–122, 148
Sandy, Kelly, 11, 112
San Martin, Robert, 11, 12, 119–120
Saudi Arabia, 51–52
Savitz, Maxine, 112, 119
Sawhill, John C., 40, 119
Schirmer, Kathryn P., 105
Schlesinger, James R.
  appointment of, 98–101
  and DPR, 104
  and MOPPS, 91
  resignation of, 118
  and SERI, 82–83
  and SFBP, 116
Schmitt, Harrison H., 30, 71
Schneider, Claudine, 167
School Heating Augmentation Experiment, 36–39
Science and Technology Association of the People's Republic of China, 52
Scott, Ronald D., 102
Seamans, Robert C., Jr.
  and ERDA, 56–57, 60
  resignation of, 99
  and SERI, 79, 81–82
Second law analysis, 193–194
Security issues vs. environmental, 176
SEGS I plant, 222
SEGS IX plant, 222
SEGS X plant, 234–235
SEGS XII plant, 223
Sensible storage, 203
Sharp, Philip, 167
Sherman, Joseph, 55
Smith, Gorman C., 41
Societal costs, 238–244
Solar 80 program, 151
Solar and Energy Conservation Exhibit, 88
Solar Applications for Buildings, 119, 121
Solar Applications for Industry, 119
Solar architecture and planning, 220
Solar augmented heat pumps (SAHP), 219
Solar Building Architecture, 206–207
Solar domestic hot water (SOHW), 215
Solar electric generating systems (SEGS), 221–225
Solar Energy Act (PL 94-473), 182
Solar Energy: A Status Report, 104
Solar Energy Coordinating Committee, 24

Index   277

Solar Energy Development Bank
 attempt to close, 144
 in Domestic Policy Review, 107
 proposed, 118
 in Reagan administration, 152–155
Solar Energy Government Buildings
 Program (SEGBP), 71
Solar Energy Industry Association (SEIA)
 codes and standards by, 133–135
 tax credit support by, 131–133
Solar Energy Information Data Bank
 (SEIDB), 125, 150
Solar Energy Laboratory, 51
Solar Energy Panel, 27–28
Solar Energy: Progress and Promise, 104
Solar Energy Research, Development and
 Demonstration Act (PL 93-473), 45, 62,
 67, 78
Solar Energy Research and Education
 Foundation (SEREF), 133
Solar Energy Research Institute (SERI)
 creation of, 78–83
 Hubbard appointment, 149–150
 management of, 130–131
 in Reagan administration, 147–150
 role of, 124–128
*Solar Energy Research Institute and
 Regional Solar Energy Centers* report,
 130–131
Solar Energy School Heating Augmentation
 Experiment, 36–39
Solar Energy Task Force, 39, 42–43, 89
Solar Federal Buildings Program (SFBP),
 71, 116–117, 228
Solar Heating and Cooling Demonstration
 Act (PL 93-409), 45, 55, 67, 71
Solar Heating and Cooling of Buildings
 program, 57
Solar Heating and Cooling Programme, 49
"Solar Heating and Cooling Systems in
 Buildings" study, 48
Solar Institute Project Office (SIPO), 79,
 81–82
Solar One power plant, 225, 228
Solar orphan systems, 170
Solar Power Applications, 119
Solar radiation and meterological database
 (SOLMET) program
 stations, 182, 184
Solar Rating and Certification Corporation
 (SRCC), 134, 194–195
Solar Resource Advisory Panel, 151
Solar resource analysis, 181–184
Solar satellite power system (SPS), 31

Solar school projects, 36–39
Solar Schools Proof of Concept Experiments, 35
Solar Standards Steering and Oversight
 Committee, 133
Solar Thermal Energy Conversion Program
 Documentation Project, 17
 approach in, 20–22
 objectives of, 18–20
SOLERAS program, 51–52
Source Evaluation Board (SEB), 81–82
Southern California Edison (SCE), 221, 225
Southern Solar Energy Center (SSEC), 140,
 151
Space cooling and heating systems, 219–220
Space heating symposium, 25
Space heating systems, 217–219
Standards, 229–230
 for collectors, 194–195
 by SEIA, 133–135
Starbird, Alfred D., 85
Stelson, Thomas, 119, 130
Stever, H. Guyford, 31, 49, 56–57
Stockman, David
 budget cuts by, 144
 and R&D, 142
 and Solar Energy Development Bank,
 153–154
Storage, 203–206
Subpanel IX report, 89, 92
Subsidies. *See* Incentives
Sununu, John, 177
Support programs, 235–236
Swimming pool heating systems, 214
Synthetic solar radiation database
 (ERSATZ), 182, 184

Taschek, Richard F., 123
Task Force on Energy, 28
Tax credits
 for domestic hot water systems, 215
 evaluation of, 232–235
 in Ford administration, 89
 in Reagan administration, 168–170
 SEIA support for, 131–133
 for space heating systems, 217
Taylor, Sam, 131
Teague, Olin E., 57, 66
Technical information programs, 230–231
Technology, 186–189
 collectors, 191–203
 energy storage, 203–206
 materials, 187, 190–191
 transfer programs, 231

Technology Development division, 126
Technology Dissemination division, 126
Teem, John M.
 at ERDA, 56–59, 62
 and ERDA and NASA roles, 71, 78
 resignation of, 84
 and SERI, 78–79, 81
Telkes, Maria, 26
Teller, Edward, 34
Tests for collectors, 194–195
Thermal storage, 203–206
Thorne, Robert D., 101, 129
Timonium Elementary School, 36–37
Training programs, 135, 230–231
Tribble, Joseph, 11, 149
 appointment of, 145–146
 departure of, 156–158
TRW (Thompson, Ramo, Woolridge, Inc.), 35
Typical meteorological year (TMY), 182–183

Union of Soviet Socialist Republics (USSR), 49–51
United Nations Educational, Scientific, and Cultural Organization (UNESCO), 48
U.S. Geological Survey (USGS), 15
Utility Photovoltaic Group demonstration program, 250

Veigel, John, 128

Walden, Omi, 11
 in advisor position, 118–119
 appointment of, 101, 110–111, 129
 commercialization and market development under, 113–116
 LASL action protested by, 123
 organization changes by, 111–113
 and RSECs, 129
Warfield, George, 126
Watkins, James D., 177
West, Ronald E., 185, 226–227
Western Regional Group, 128
Western Solar Utilization Network (Western SUN), 151
Westinghouse, 35, 52
White, Philip C., 91
Winston, Roland, 193
Wirth, Tim, 167, 179
World Symposium on Applied Solar Energy, 25

Zarb, Frank G., 56–57, 66
Zero-based budgeting (ZBB), 102
Zero energy growth (ZEG) scenario, 53–54